Spanning Organisational and Geographical Boundaries

AF694494

Michael Mayer

Spanning Organisational and Geographical Boundaries

Understanding Supply Chain Risk Management in Hidden Champions

Springer Gabler

Michael Mayer
Fachbereich Wirtschaft
FH Vorarlberg
Dornbirn, Austria

This text is a reprint of a dissertation submitted to Bournemouth University in 2023 entitled 'Spanning organisational and geographical boundaries: Understanding Supply Chain Risk Management in Hidden Champions'.
ID Code: 38173

ISBN 978-3-658-50452-6 ISBN 978-3-658-50453-3 (eBook)
https://doi.org/10.1007/978-3-658-50453-3

© The Editor(s) (if applicable) and The Author(s), under exclusive license to Springer Fachmedien Wiesbaden GmbH, part of Springer Nature 2026

This work is subject to copyright. All rights are solely and exclusively licensed by the Publisher, whether the whole or part of the material is concerned, specifically the rights of translation, reprinting, reuse of illustrations, recitation, broadcasting, reproduction on microfilms or in any other physical way, and transmission or information storage and retrieval, electronic adaptation, computer software, or by similar or dissimilar methodology now known or hereafter developed.
The use of general descriptive names, registered names, trademarks, service marks, etc. in this publication does not imply, even in the absence of a specific statement, that such names are exempt from the relevant protective laws and regulations and therefore free for general use.
The publisher, the authors and the editors are safe to assume that the advice and information in this book are believed to be true and accurate at the date of publication. Neither the publisher nor the authors or the editors give a warranty, expressed or implied, with respect to the material contained herein or for any errors or omissions that may have been made. The publisher remains neutral with regard to jurisdictional claims in published maps and institutional affiliations.

This Springer Gabler imprint is published by the registered company Springer Fachmedien Wiesbaden GmbH, part of Springer Nature.
The registered company address is: Abraham-Lincoln-Str. 46, 65189 Wiesbaden, Germany

If disposing of this product, please recycle the paper.

This book is dedicated to my wife Alexandra and my daughters Rosa Maria, Clara, and Emilia and to my son Hannes. Thank you for supporting, tolerating, and encouraging me throughout this research project.
Finally, If I could, I would like to thank my late father, Friedemann H. Mayer, for his trust and support.

Foreword

As one of Michael's doctoral supervisors, I am honoured to write the foreword for this publication. He showed immense patience as a doctoral student and allowed me and my colleague, Dr Tania Humphries-Smith (co-supervisor), to be meticulous in our attention to detail, and in challenging his thinking. Michael is someone who thinks twice and is not afraid not to be part of a consensus.

We have shared many thought-provoking in-depth discussions, but it is clear that Michael is an operator, not an armchair philosopher. Michael thinks from first principles, tests things and always questions. He is prepared to change his mind, and is not good at fooling himself. He thinks holistically, but doesn't take himself too seriously—an important attribute that many could benefit from sharing!

This book provides an understanding of Supply Chain Risk Management in the mechanical engineering industry, in particular within organisations that provide a substantial contribution to GDP. These are often referred to as Hidden Champions in regions of Germany, Switzerland and Austria. The importance of the research reported in this book has been emphasised by the global Covid 19 pandemic, in which many supply chains were evidently shown not to be resilient.

This book reports on ways to increase supply chain resilience and considers Hidden Champion Supply Chain Risk Management processes with those of their supply chain partners.

Prof. Dr. Clive Hunt
Associate Professor
Bournemouth University

Acknowledgements This book would not have been possible without the support and cooperation of the senior managers of the companies participating in this study. They generously offered their time and experience during the interviews and the focus group discussion. Once again, thank you very much for this.

Furthermore, I would like to thank like to thank Prof. Dr. Clive Hunt and Prof. Dr. Tania Humphries-Smith for their time, committed support and guidance throughout my research journey. Without your challenging questions and the lively discussions we had, I would not have learned that much! Thanks also to the faculty and academic staff of Bournemouth University, Naomi Bailey, the IT and library team who have supported me on my journey to achieving this milestone.

Next to my supervisors, I would also like to thank the members of my Viva Voce committee: Prof. Dr. Mehran Koohgilani, Prof. Dr. Lydia Bals, Prof. Dr. Diogo Montalvão. It is a great honour for me that you were part of my committee and showed your interest in my research topic. I appreciated your valuable comments, which helped me to improve the quality of my doctoral thesis.

I would also like to thank those who took the first steps on this path together with me. The faculty at what is now the Technical University of Applied Sciences Würzburg-Schweinfurt, especially Prof. Dr. Karsten Machholz. With Karsten, I was able to hold my first academic position and with him, I was allowed to represent the Technical University of Applied Sciences Würzburg-Schweinfurt in Saudi Arabia. Also, Prof. Dr. Marcus Schulz gave me a new perspective on our way up the "Holy Mountain of Franconia". I hope the three of us will be able to make the way up the holy mountain together some day. I would also like to thank the faculty of the Duale Hochschule Baden-Württemberg in Ravensburg, Heidenheim and Stuttgart as well as the faculty at Kempten University of Applied Sciences.

Finally, I would like to thank my friends and colleagues who have always given me advice when asked for.

Michael H. Mayer

Competing Interests The author has no competing interests to declare that are relevant to the content of this manuscript.

Contents

1 "Disruptive Events Challenge Hidden Champions' Global Supply Chains" 1
1.1 Introduction 1
1.2 Aims and Objectives of this Thesis 3
1.3 Outline of this Thesis 4

2 Review of Extant Literature in the Constituting Research Elements 7
2.1 Approach to Reviewing the Literature 7
2.2 Structure of this Literature Review 8
2.3 Unique Features of Hidden Champions 9
2.3.1 Significance of Hidden Champions for the Economy 10
2.3.2 Geographic Concentration of Hidden Champions 12
2.3.3 Success Strategies of Hidden Champions 14
2.3.4 Vulnerability of Global Supply Chains 18
2.3.5 Regulatory Framework for Supply Chain Risk Management 19
2.4 Supply Chain Risk Management and Hidden Champions 20
2.4.1 Resilience of Hidden Champions 21
2.4.2 Ambiguity of Supply Chain Risk Management Definitions 21
2.4.3 Differing Scope of the Underlying Supply Chain 23
2.4.4 Differing Schools of Supply Chain Risk Management Processes 23

2.4.5 Definition of Supply Chain Risk Management 25
2.4.6 Supply Chain Risk Management Research in Hidden Champions 27
2.4.7 Small and Medium-Sized Enterprise Perspective 27
2.4.8 Family Business Perspective 28
2.4.9 Theoretical Perspectives 29
2.5 Gaps in the Literature 30
2.6 Identifying Areas for further Research 31
2.6.1 Supply Chain Risk Management Cooperation 32
2.6.2 Supply Chain Risk Management Training 33
2.7 Research Question 34
2.8 Chapter Summary 34

3 Theoretical Framework of this Research 37
3.1 Prevailing Theories in the Reviewed Literature 38
3.2 Supply Chain Practice View: a Theoretical Framework 39
3.3 Interorganisational Practices 41
3.4 Chapter Summary 44

4 Research Methodology 45
4.1 Research Philosophy 47
4.1.1 Ontological Assumption 49
4.1.2 Epistemological Assumption 49
4.1.3 Axiological Assumption 50
4.1.4 Interpretive Approach to Research 50
4.2 Research Design 51
4.3 Research Ethics 53
4.4 Research Methods 54
4.4.1 Reflection and Learning from the Pilot Phase 54
4.4.2 Purposeful Sampling Strategy 55
4.4.3 Data Collection using Semi-structured Interviews 57
4.4.4 Coding and Data Analysis 59
4.4.5 Transcription and Translation 60
4.4.6 Anonymisation of Interview Partners 64
4.4.7 Data Structure 66
4.4.8 Validate Findings through Focus Group Discussion 67
4.5 Challenges and Limitations of this Research Design 70
4.6 Chapter Summary 72

5 Managing Risks in the Hidden Champions' Supply Chains 73
5.1 Findings and Discussion from the Semi-structured Interviews 74
5.2 Supply Chain Risk Management Process 74
5.2.1 Supply Chain Risk Management Process Ownership 75
5.2.2 Supply Chains with limited Complexity 76
5.2.3 Complex Upstream Supply Chains 76
5.2.4 Supply Chain Risk Management Designation 79
5.2.5 Supply Chain Risk Management Process Definition 80
5.2.6 Supply Chain Risk Management Approach 82
5.2.7 Supply Chain Risk Management Performance Indicators 83
5.2.8 Supply Chain Risk Management Process Digitalisation 85
5.3 Supply Chain Risk Management Culture 87
5.3.1 Supply Chain Risk Management Awareness 88
5.3.2 Supply Chain Risk Management Training 89
5.3.3 Supply Chain Risk Management Talent Management 92
5.4 Supply Chain Risk Management Cooperation 94
5.4.1 Close Ties in the Market Segment 95
5.4.2 Long-term Partnerships 96
5.4.3 Interorganisational Cooperation 97
5.4.4 Interorganisational Supply Chain Risk Management Training 104
5.4.5 Supply Chain Risk Management Process Audit 105
5.4.6 Supply Chain Risk Management Development 107
5.4.7 Supply Chain Risk Management Process Harmonisation 108
5.4.8 Summary of the Findings from the Semi-structured Interviews 109
5.5 Findings and Discussion from the Focus Group Discussion 110
5.5.1 Cross-Sectoral Sampling Strategy 111
5.5.2 Discussing Supply Chain Risk Management Processes 111
5.5.3 Unpronounced Supply Chain Risk Management Culture 112

5.5.4 Isolated Supply Chain Risk Management 114
5.5.5 Summary of the Findings from the Focus Group Discussion 117
5.6 Chapter Summary 119

6 Supply Chain Risk Management—A Cooperative Approach 121
6.1 Interorganisational View of Supply Chain Risk Management 122
6.2 Adopting a Cooperative Supply Chain Risk Management Approach 125
6.3 Process Standardisation and Improvement 126
6.3.1 Two-Phase Model Process Model 127
6.3.2 Process Standardisation Cycle 128
6.3.3 Process Improvement Cycle 131
6.4 Discussing the Emergent Framework 134
6.4.1 Formalised Supply Chain Risk Management Processes 134
6.4.2 Corporate Supply Chain Risk Management Culture 135
6.4.3 Cooperative Supply Chain Risk Management 136
6.5 Chapter Summary 137

7 Conclusion and Recommendations for Future Work 139
7.1 Findings in Relation to the Research Questions 139
7.2 Personal Reflection 142
7.3 Contribution to Knowledge 143
7.3.1 Contribution to Theory 144
7.3.2 Contribution to Practice 145
7.4 Limitations and Recommendations for Future Research 147
7.4.1 Limitations of this Research 148
7.4.2 Recommendations for Future Research 148

References 151

About the Author

Michael Mayer serves as Lecturer for Supply Chain and Operations Management at the Vorarlberg University of Applied Sciences, Dornbirn in Austria and is the founder and owner of RMB Beratung. Based on the topic of this Thesis, he is offering Training and Consultant Work for small and medium enterprises on Supply Chain Risk Management and the German Supply Chain Act (Lieferkettengesetz) and the European Supply Chain Due Dilligence Directive.

Prior to this, Dr. Michael Mayer served as a purchasing and supply chain executive with over 25 years of international Purchasing, Supply Chain, and Quality experience within the automotive and raw material industries. He successfully restructured purchasing and supply chain organisations, implementing best practice methodologies and systems to improve financial results and successfully driving turnarounds in the business environment.

Dr Mayer speaks fluent German and English and has a good knowledge of French and Italian. He holds a doctorate from Bournemouth University, UK, and a degree in industrial engineering from the Technical University of Würzburg-Schweinfurt.

Abbreviations

ASME	Austrian Institute for SME Research
BSR	Buyer-Supplier Relationship
CEE	Central and Eastern Europe
DACH region	German-speaking region of Germany, Austria and Switzerland
DIY	Do-It-Yourself
EC	European Commission
EDI	Electronic Data Interchange
ERM	Enterprise Risk Management
ERP	Enterprise Resource Planning
ESI	Early Supplier Involvement
EU	European Union
FGD	Focus Group Discussion
FMEA	Failure Mode and Effect Analysis
FMTI	Fachverband Metalltechnische Industrie (Metaltechnology Austria)
GDP	Gross Domestic Product
HC	Hidden Champion
HRM	Human Resource Management
IFM	Institut für Mittelstandsforschung Bonn (Institute for SME-Research Bonn, Germany)
IRÄG	Insolvenzrechtsaenderungsgesetz (Austrian Insolvency Amendment Act)
KonTraG	Gesetz zur Kontrolle und Transparenz im Unternehmensbereich (German Corporate Control and Transparency Act)

KPI	Key Performance Indicator
KUA	Kurzarbeit (Short-time Work)
MRP	Materials Requirement Planning
NPD	New Product Development
OR	Obligationsrecht (Swiss Code of Obligations)
OTIF	On Time-In Full
PBV	Practice-based View
PDCA Cycle	Plan-Do-Check-Act Cycle
PPM	Parts Per Million
RBV	Resource-Based View
REC	Bournemouth University Research Ethics Committee
RM	Risk Management
RV	Relational View
SC	Supply Chain
SC partner	Supply Chain Partner
SCM	Supply Chain Management
SCO	Swiss Code of Obligations
SCPV	Supply Chain Practice View
SCRM	Supply Chain Risk Management
SEW	Socioemotional Wealth
SME	Small and Medium-Sized Enterprises
SME Portal	SME Portal of Switzerland
SWISSMEM	Swiss Association of Mechanical and Electrical Engineering Industries
VMI	Vendor Managed Inventory

List of Figures

Fig. 1.1 Thesis Structure 4
Fig. 2.1 Key Figures of the Mechanical Engineering Industry Sector (FMTI 2021; SWISSMEM 2021; VDMA 2021) 11
Fig. 2.2 Comparison of Export Volumes (SWISSMEM 2021) 12
Fig. 2.3 Number of Hidden Champions per 100,000 Capita 13
Fig. 2.4 Hidden Champions Model 15
Fig. 2.5 Visualisation of Supply Chain Risk Management Definitions 26
Fig. 3.1 Theories influencing the Supply Chain Practice View 40
Fig. 3.2 Framework of this Research 42
Fig. 4.1 Structure of Chapter 4: Research Methodology 46
Fig. 4.2 Continuum of the Research Paradigms 50
Fig. 4.3 Example of Interview Question and Coding 62
Fig. 4.4 Example of Smoothened Coding 62
Fig. 4.5 Example Focus Group Discussion Coding 64
Fig. 4.6 Data Structure of this Research 66
Fig. 5.1 Linear Upstream Supply Chain 76
Fig. 5.2 Complex Upstream Supply Chain 78
Fig. 6.1 Emergent Theoretical Framework of this Research 124
Fig. 6.2 Plan-Do-Check-Act Cycle 126
Fig. 6.3 Process Standardisation Cycle 129
Fig. 6.4 Process Improvement Cycle 132

List of Tables

Table 2.1 Legal Requirements for Enterprise Risk Management 20
Table 2.2 Most cited Supply Chain Risk Management Definitions 22
Table 2.3 Key Elements of the Christopher-School 24
Table 2.4 Key Elements of the Norrman-School 24
Table 2.5 Key Elements of the Pfohl-School 25
Table 4.1 Research Philosophies 48
Table 4.2 Excerpt of Potential Interview Contacts 56
Table 4.3 Timeline of this Research 59
Table 4.4 Excerpt Coding Table 63
Table 4.5 Interview Pseudonyms 65

Table 2.1 Legal Requirements for Business Risk Management ... 20
Table 2.2 Normative Supply Chain Risk Management Definitions ... 22
Table 2.3 Key Elements of the Classification-Based Method ... 24
Table 2.4 Key Elements of the Taxonomy Method ... [illegible]
Table 2.5 Key Elements of the Mixed-Method ... [illegible]
Table 3.1 Research Preferences ... [illegible]
Table 3.2 Overview of Potential Interviewed Industry Experts ... [illegible]
Table 4.1 Timeline of this Research ... 59
Table 4.2 Conceptual [illegible] ... [illegible]
Table 4.3 Timeline of Research Years ... 65

1 "Disruptive Events Challenge Hidden Champions' Global Supply Chains"

Disruptive events threaten the viability (Ivanov 2021) of global supply chains (SC). According to Stevens (1989), supply chains are a set of activities, including planning, coordinating, and controlling the flows of materials and information from the source of supply to the point of consumption that contribute to value creation by means of upstream and downstream connections. This thesis focuses on disruptive events emerging in the upstream supply chains of Hidden Champions (HC), niche-leading and mainly small and medium-sized family businesses (Simon 1990). These upstream supply chains span from the Hidden Champions to the supply chain partners (SC partner) that are in a direct business partnership with the Hidden Champions. Given their elevated levels of globalisation, the Hidden Champions have increasingly complex supply chains. Along with their global focus, which is typical for niche-leading businesses to generate growth, it is also the scarcity of skilled employees that is typical for the Hidden Champions. This scarcity of skilled employees was brought forward in a notable quote of Simon, stating that Hidden Champions had *"More work than people"* (2009, p. 268), which has become even more prevalent in recent years (Simon 2020a). Despite these challenges the Hidden Champions are experiencing, they enjoy a high resilience against disruptive events (Simon 2017), making them an exciting object of study in the context of this thesis.

1.1 Introduction

Examples of supply chain disruptions are manifold. Be it the COVID-19 pandemic that has unprecedentedly challenged supply chains globally (Ivanov 2021) or the ongoing energy crisis that since 2020 leads to soaring prices for gas and electric energy having a direct impact on consumers worldwide and forecasted

© The Author(s), under exclusive license to Springer Fachmedien Wiesbaden GmbH, part of Springer Nature 2026

M. Mayer, *Spanning Organisational and Geographical Boundaries*,
https://doi.org/10.1007/978-3-658-50453-3_1

by *Deutsche Welle* (Pandey 2021) to continue to hold back the recovery of global industry production. One further example of an upstream supply chain disruption is the case of Ericsson, a former Swedish mobile phone producer, well-discussed in *The Wall Street Journal* (Latour 2001) or by Norrman and Jansson (2004); Chopra and Sodhi (2004); Blome and Henke (2009) or Sodhi and Tang (2012). Latour (2001) reported that Ericsson lost over $400 million after its upstream supply chain was disrupted by a fire at a supply chain partner. This case has recently gained increasing popularity again among practitioners due to its similarity with the global shortage of semiconductors in the aftermath of the COVID-19 pandemic. According to *The Guardian* (Partridge 2021), the semiconductor shortage until June 2021 caused a loss of $20 billion in global motor vehicle production alone and has a rippling effect on related industry sectors such as the mechanical engineering industry (Ivanov et al. 2014) threatening business continuity in those related sectors. Given this threat to business continuity, the author sought for companies with a high significance for the economy and to identify their approach to managing and mitigating disruptive events in their upstream supply chain to avoid losses or business closure. Hidden Champions, defined as global or regional market leaders in their segment; with annual revenue below €5 billion and a low level of public awareness that is not further quantified according to Simon (2021), have this high significance for the economy since they generate gross domestic product (GDP) and employment (Simon 2017). Based on their global success, the Hidden Champions' supply chains have grown global and therefore are increasingly complex and are prone to disruptive events. The management and mitigation of such disruptive events in supply chains are called *Supply Chain Risk Management* (SCRM). Research on Supply Chain Risk Management is in the early stages of development and has received increasing attention from both academics and practitioners since the beginning of the new millennium triggered by numerous supply chain disruptions, threatening companies' survival (Zsidisin 2019). Within the cross-border area of the German-speaking region of Germany, Austria and Switzerland, a particular high geographic concentration of Hidden Champions exists (Simon 2021). Therefore, this study will focus on Hidden Champions in the cross-border area of the so-called DACH region. The acronym DACH is the combination of the international Vehicle Registration Codes of the three neighbouring countries Germany (D), Austria (A) and Switzerland (CH). Research on Supply Chain Risk Management in Hidden Champions in the cross-border area of the DACH region does not exist. The focus of risk management in Hidden Champions lies in managing financial risks, as introduced in the works of Huber and Gith (2009); Lebedev (2013) or Mietzner et al. (2018), which further motivates and defends this study.

1.2 Aims and Objectives of this Thesis

This study aims at understanding and explaining how interorganisational cooperation requiring training and development of employees between Hidden Champions and their supply chain partners is mediating or positively influencing (OLD 2022), efficient Supply Chain Risk Management. In the case of supply chain disruptions, efficient management and mitigation of disruptive events require understanding the Hidden Champions' Supply Chain Risk Management processes and swift and seamless communication to ensure business continuity.

The following five research objectives address this aim:

- To understand the Supply Chain Risk Management processes used in Hidden Champions in the DACH region;
- To identify education and experience along with the scope of responsibility of the Supply Chain Risk Management process owner in Hidden Champions;
- To understand training and development provided to Hidden Champions' employees to apply the Supply Chain Risk Management processes;
- To understand the communication among the members of the Hidden Champions' upstream supply chain to mediate the application of Supply Chain Risk Management processes;
- To understand the training and development provided to Hidden Champions' supply chain partners to support the Hidden Champions' Supply Chain Risk Management processes.

By meeting the above aim and objectives, this study explains how interorganisational cooperation between Hidden Champions and their supply chain partners is mediating the managing and mitigating disruptions in the Hidden Champions upstream supply chain. Training the Hidden Champions' employees and those of their supply chain partners can create mutual understanding for the importance of Supply Chain Risk Management to manage and mitigate disruptive events in the Hidden Champions' upstream supply chain. Successful cooperation among the members of the supply chain, defined as supply chain members (Wang et al. 2008), can lead to superior performance of the Supply Chain Risk Management processes, which should ensure business continuity and identifies business opportunities ahead of the Hidden Champions' competitors.

1.3 Outline of this Thesis

This section presents the outline of this thesis, that is structured in a linear manner (Fig. 1.1).

Fig. 1.1 Thesis Structure

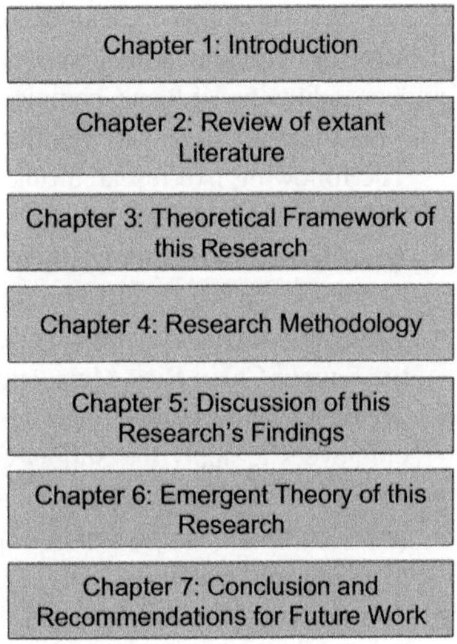

Chapter 1 introduces the background of this study by justifying the need to manage and mitigate disruptive events in supply chains threatening the business continuity of Hidden Champions, largely unknown, globally successful niche-leading mainly small and medium-sized family businesses.

The literature review presented in Chapter 2 is structured logically and organised according to the two constituting elements of this research project, the Hidden Champions and Supply Chain Risk Management. The review first offers an in-depth discussion of the Hidden Champions' characteristics and their significance for the economy. The relevance of Supply Chain Risk Management at the Hidden Champions based on their global and complex supply chain is provided in the second section. With the combination of both research areas, Hidden Champions and Supply Chain Risk Management, the gaps identified in the literature will be discussed to define areas for further research. The concluding section of this chapter points out the need for research on cooperation in a Supply Chain Risk Management context and identifies the research question.

Based on aims, objectives, and this study's research question, the theoretical framework for this research is introduced in *Chapter 3* of this thesis. Identifying a theoretical lens suitable for this research supports the development of a conceptual framework.

Chapter 4 introduces, justifies, and explains the methodological choices for the inquiry into qualitative data answering the previously defined research question. Following the research design, the relevant ethical considerations are discussed. The applied research methods for collecting, coding and analysing qualitative data, validating the research findings, and finally, the challenges and limitations experienced based on this research design are reported to provide qualitative rigour, delivering plausible and defensible conclusions.

Chapter 5 builds on the translated narratives giving extraordinary voice to the interview partners of this study and provides an informative discussion of this research's findings on intraorganisational Supply Chain Risk Management in Hidden Champions in light of the existing literature.

Chapter 6 introduces the emerging theory from this research project. Based on the data from the previous discussion of the findings, the concept of cooperative Supply Chain Risk Management supports an interorganisational view spanning organisational boundaries of the Hidden Champions to manage better and mitigate the disruptive events in the upstream supply chain.

This thesis concludes with a summary in *Chapter 7* reflecting on the experiences from the research process and presenting the contribution to knowledge from a theoretical and a practical perspective as a principal component. Furthermore, this chapter reflects upon the limitations of this research and provides recommendations for further research.

Review of Extant Literature in the Constituting Research Elements

2

Hidden Champions are of high relevance for the economy in the DACH region (Landau et al. 2016); their increasing globalisation patterns result in complex supply chains (Tang 2006b). Research by Christopher and Lee (2004) and Jüttner (2005) contemplates that those complex supply chains are prone to increased vulnerability. Hence, this study investigates the mediating role of interorganisational cooperation between Hidden Champions in the DACH region and their upstream supply chain partners in managing and mitigating disruptive events. This study aims to understand and explain the Supply Chain Risk Management processes used by Hidden Champions and identify the education and experience of the Hidden Champions' employees that are applying the Supply Chain Risk Management processes. A further objective of this study is to understand the communication among the supply chain members to mediate or positively influence the management and mitigation of disruptive supply chain events. Building on this communication, a further objective is to understand the training and development that Hidden Champions provide to their supply chain partners' employees to support the Supply Chain Risk Management processes.

2.1 Approach to Reviewing the Literature

This initial review covers the two constituting elements of this research: the Hidden Champions and Supply Chain Risk Management. Hence it provides an overview of the literature available for Supply Chain Risk Management in Hidden Champions in the cross-border area of the German-speaking federal states and cantons of the DACH region. The subsequent sections review and discuss the Hidden Champions' unique features, including their economic significance. Moreover, the subsequent sections discuss the relevance of Supply Chain Risk

© The Author(s), under exclusive license to Springer Fachmedien Wiesbaden GmbH, part of Springer Nature 2026

M. Mayer, *Spanning Organisational and Geographical Boundaries*,
https://doi.org/10.1007/978-3-658-50453-3_2

Management within Hidden Champions based on their global and complex supply chains. This discussion introduces prevailing theoretical perspectives in the reviewed literature on Hidden Champions. The review concludes by identifying the gaps in the literature on a topic area that is widely understudied, bearing significant insufficiencies in research, both in a thematic context and theoretical perspective. The gaps identified in the literature review propose areas for further research by applying Supply Chain Risk Management across organisational borders and generating future knowledge in Hidden Champions and their supply chain partners. Literature reviews in theses using the Gioia Method (Gioia et al. 2013) should always consider an enforced ignorance of the existing research concepts, which allows a maximum openness of the researcher to identify new concepts rather than being blindfolded by the existing concepts (Gioia et al. 2013; Gehman et al. 2018). However, an initial review of the relevant literature was performed to identify the research gaps in the fields of Supply Chain Risk Management in Hidden Champions to develop a research question for this research. The approach used to initially identify, review, and analyse the literature was inspired by the systematic review process introduced by Tranfield et al. (2003). A set of inclusion criteria was defined to identify studies and to avoid bias by purely selecting literature based on subjective criteria. These inclusion criteria also ensured that only topic-relevant documents were identified for this review proposed by (Tranfield et al. 2003).

2.2 Structure of this Literature Review

This literature review is structured according to the two constituting elements of this research, Hidden Champions and Supply Chain Risk Management. The following sections first offer a discussion of the Hidden Champions' unique features, their significance for the economy and their geographic concentration in the DACH region. Success strategies of the Hidden Champions demonstrate their need for globalisation that results in the vulnerability of their supply chains; hence regulatory frameworks for risk management and supply chain risk management are required to reduce the loss for the single business or the overall economic loss. The discussion on the Hidden Champions' unique features is followed by insights provided on Supply Chain Risk Management research, which identifies ambiguous definitions for Supply Chain Risk Management itself, a differing scope of the underlying supply chain and differing schools of Supply Chain Risk Management processes. With no dedicated research on Supply Chain Risk Management in the existing literature on Hidden Champions, this review expands to small and medium-sized business and family business literature. In contrast, both research streams find entrance in Supply Chain Risk Management and Hidden Champions research. With the research on Hidden Champions, family businesses, small and medium-sized enterprises (SME) and Supply Chain Risk Management, this review identifies different theoretical perspectives that distinguish different areas for research that are discussed in the next section of this review. The initial review of the existing literature in the thematic areas of Hidden Champions and Supply Chain Risk Management provides rationale and context for this study, supporting sensitivity and critical thinking on the existing concepts in these areas as proposed by Dunne (2011). Based on this initial literature review, this thesis identifies the research gaps of cooperation, training and development of employees in Supply Chain Risk Management in Hidden Champions. These research gaps then give rise to the topics for further work conducted in this thesis. The concluding section of this literature review points out the need for a research question on interorganisational cooperation in a Supply Chain Risk Management context.

2.3 Unique Features of Hidden Champions

Hidden Champions are a global phenomenon and not restricted to Germany. Simon (1990) initially coined the term Hidden Champions, derived from globalisation and export success of the German-speaking and Scandinavian countries with Theodore Levitt at Harvard Business School in the late 1980s following Levitt's publication (1983) on the globalisation patterns of multinational companies. According to Simon's (2021) most recent definition, Hidden Champions are global or regional market leaders in their segment; their annual revenue is below €5 billion and there is a low level of public awareness that is not quantified. This low level of public awareness, however, is mainly due to family ownership structures and absence from the stock market, although Simon (2009) declares that there is an increasing number of Hidden Champions that have tried to obtain access to the stock market.

A unique feature of Hidden Champions is their financial independence from the organised capital market that enables long-term strategic planning and growth rather than quarterly reporting within public companies. Their strong bonds within the region where the business resides are very typical. As a result, Hidden Champions are perceived as trustworthy business partners by their stakeholders: their customers, the principal bank, often the Hidden Champions' only banking partner known as the "Hausbank" (Lehrer and Schmid 2015, p. 305), and for their employees, who very often live in close vicinity to the factory building. Simon (2009) underpins the Hidden Champions trustworthiness by traditional values such as reputation, reliability and distinct risk aversion. Also, the managerial behaviour in these companies follows traditional values, based on a patriarchal structure that is epitomic for many Hidden Champions in the cross-border area of the DACH region. At many of the Hidden Champions, the owner-manager of the business is called the Patriarch or the Patron, who is often 100 per cent personally liable for the business (Lehrer and Schmid 2015). The Patron is a respected person in society and often has strong ties to the political decision-makers on regional and on a national level as reported by *Süddeutsche Zeitung* (Busse and Mayr 2019) in an article on long-standing family firms in the German fashion industry. Due to the export success patterns of the Hidden Champions, they enjoy high international interest as a model to follow. However, empirical research remains scarce (Schlepphorst et al. 2016). The majority of research has been published by Simon (Cf. Simon 1990; 1992; 1996; 2009; 2017, 2019, 2020a, b; 2021). Simon's research has a strong focus on the success strategies of the Hidden Champions, isolating niche specialisation, innovation, globalisation, employee training, and leadership culture (Simon 1996).

2.3.1 Significance of Hidden Champions for the Economy

While no specific data regarding their significance for the economy in the DACH region exists, the organisations constituting the Hidden Champions, small and medium-sized enterprises, and family businesses demonstrate this significance. At the same time, they play an essential role in global and regional economic development. Small and medium-sized enterprises and family businesses employ the majority of the workforce and offer a considerable share of apprenticeships. Therefore, they actively contribute to employment and education identified by Meyer-Stamer and Wältring (2000) or published by official institutions focusing on small and medium-sized enterprises in the countries of the DACH region such as the Austrian Institute for SME Research (ASME), Vienna (2022); Institut für Mittelstandsforschung (IFM) Bonn (2022); SME Portal of Switzerland (SME Portal), Bern (2022). By contributing significantly to gross domestic product generation (Ayyagari et al. 2007) in their respective economies in the DACH region (Landau et al. 2016), the Hidden Champions play an essential role in the global and regional economy. See Fig. 2.1, that provides an overview of the 2019 key figures of the mechanical engineering industry in the three countries in the cross-border area of the DACH region.

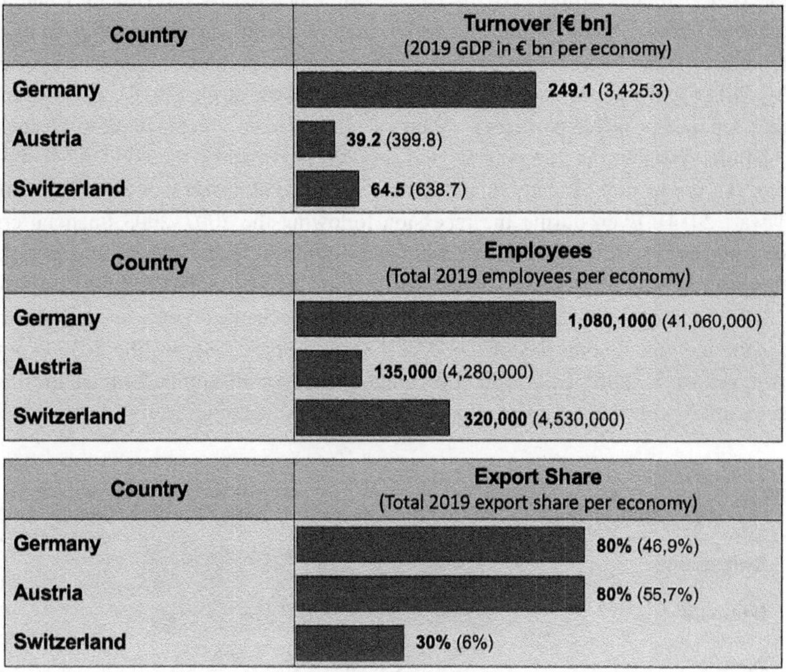

Fig. 2.1 Key Figures of the Mechanical Engineering Industry Sector (FMTI 2021; SWISSMEM 2021; VDMA 2021)

Being among the most successful and export-strongest industries in the respective economies, the mechanical engineering industry also has a strong representation with political decision-makers through their respective associations, the German Mechanical Engineering Industry Association (VDMA), the Austrian Association of Metal Technology Industries (FMTI); the Swiss Association of Mechanical and Electrical Engineering Industries (SWISSMEM). Further, the Hidden Champions in the mechanical engineering industry represent a significant share of over 20% or more of the total population of Hidden Champions in their respective economies, as presented by Rammer and Frietsch (2015), Credit Suisse AG (2018) and Cencic (2019). The reported key figures of the Hidden Champions in the mechanical engineering industry emphasise their high relevance for the studied economies, states Simon (2017). He discusses the contribution to the gross domestic product of the Hidden Champions in the manufacturing sector in the DACH region (Simon 2017). The wealth generated by the Hidden Champions, their strong bonds within the region they reside in, and the owning family's reputation make them respected and reliable business partners which is supported by Miller and Le Breton-Miller (2014) or Audretsch et al. (2018). As a result, they are indispensable partners for their customers, mainly in business-to-business relations. Without the components and services provided by the Hidden Champions, the production of many final products would be almost impossible (Buse and Tiwari 2014). Even during the recession following the 2007–2008 financial crisis, the vast majority of Hidden Champions were reliable business partners since they were resilient and recovered swiftly from the economic downturn (Simon 2017). To further illustrate the relevance of the mechanical engineering industry for the individual countries of the DACH region, Fig. 2.2 shows the 2019 export volumes in € million per 100,000 capita in an overall comparison of the five economies with the strongest exports in the mechanical engineering industry.

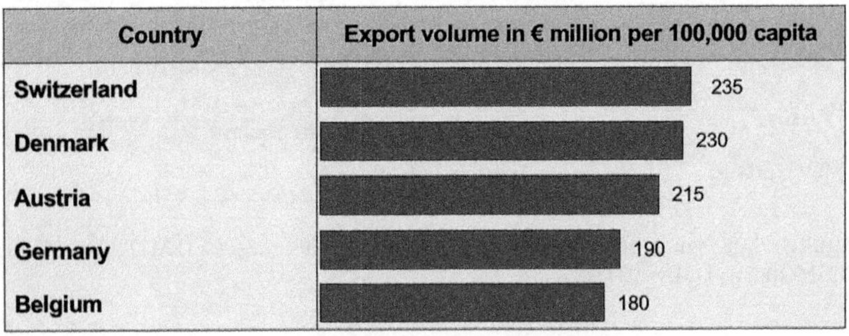

Fig. 2.2 Comparison of Export Volumes (SWISSMEM 2021)

2.3.2 Geographic Concentration of Hidden Champions

The geographic focus of this study is motivated by the high geographic concentration of Hidden Champions per capita. In his recent book, Simon (2021) identified 3,406 Hidden Champions worldwide.

An overview of the number of Hidden Champions per 100,000 capita per country portrays the three countries included in this research along with their two states or cantons with the highest per capita concentration of Hidden Champions compared to other strong economies like France, the United Kingdom of Great Britain and Northern Ireland and the United States of America (see Fig. 2.3).

Country	Hidden Champions per 100,000 capita
Switzerland	1.98
- Basle-Country	3.30
- St. Gallen	2.83
Austria	1.92
- Vorarlberg	3.30
- Upper Austria	2.91
Germany	1.89
- Baden-Württemberg	3.31
- Bavaria	2.31
France	0.17
United Kingdom	0.11
United States	0.11

Fig. 2.3 Number of Hidden Champions per 100,000 Capita. (Adapted from Simon 2021, pp. 51–59)

A more detailed view of the Hidden Champions per capita concentration in the DACH region gives insight into their exact distribution; Simon (2021) compiled their concentration in the German federal state of Baden-Württemberg, home to 367 Hidden Champions, i.e. 3.3 Hidden Champions per 100,000 inhabitants or Bavaria: 2.3 (Simon 2021). Based on Venohr and Lang (2014), the two southern districts of Baden-Württemberg, Bodenseekreis and Ravensburg, and the Bavarian district of Lindau have a specifically high concentration of around 6 Hidden

Champions per 100,000 capita. Also, the per capita concentration of Switzerland: 1.98 and Austria: 1.92 in contrast to other strong economies, i.e. France: 0.17, United Kingdom: 0.11 or the USA: 0.11 (Based on Simon 2021) and the close vicinity of the Austrian state Vorarlberg: 3.3 and the Swiss Cantons Basle-Country: 3.3 or St. Gallen: 2.8 to the districts Bodenseekreis, Lindau and Ravensburg promotes the focus of this research project on the DACH region. This high concentration of Hidden Champions in the DACH region's cross-border area justifies this study's regional focus. According to Pundy (2015), the high concentration is based on cultural similarities in this region, which is characterised by its rural territories and known to the broader public as a region for winter holidays and mountain activities with a strong sense of traditional customs and festivities. However, this region offers an intense concentration of companies with substantial export volumes supporting the significance for the economy of the Hidden Champions in the DACH region. The above discussed export volumes emphasise one of Hidden Champions' inherent strengths, namely their consistent focus on globalisation for further business growth. This continued focus of the Hidden Champions on globalising their markets for further growth is one of their success strategies which the following section will seek to understand.

2.3.3 Success Strategies of Hidden Champions

The success of the Hidden Champions has long been a topic of discussion among academics and practitioners. Their combination of narrow market focus and superior performance brands them as valued business partners when following the system of competitive strategies introduced by Michael Porter (1985). To explain the Hidden Champions' success strategies, Simon (2009) uses the Hidden Champions Model that illustrates their success strategies in three circles. The nucleus of the Hidden Champions Model is visionary leadership in combination with ambitious goals. Depth of the value chain, decentralisation of responsibilities and high performance among employees can be identified in the inner shell around the nucleus. The outer shell embraces focus by concentrating on a specific niche, globalisation for business growth, innovation activities and customer proximity which are seen as success strategies in connection with leadership styles that are authoritarian and participatory at the same time (Simon 2009). What follows is a discussion that introduces the success strategies identified in this review and discusses their mediating influence on managing and mitigating disruptive events in the Hidden Champions' upstream supply chain, (Fig. 2.4).

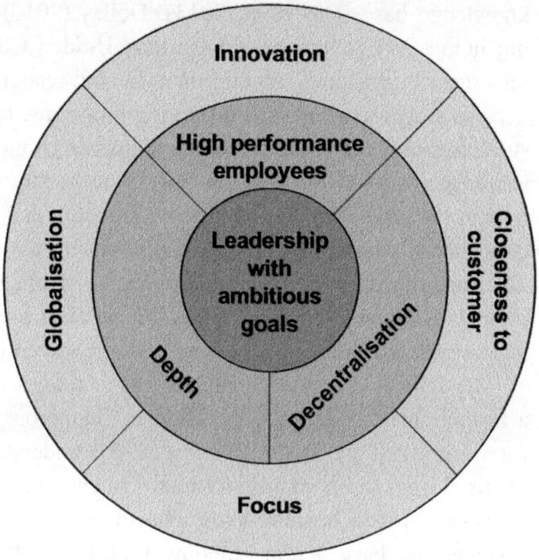

Fig. 2.4 Hidden Champions Model. (Adapted from Simon 2009, p. 356)

The most prominent success strategy identified is innovation, creating value both in production technology and the development of unique products. Vincent et al. (2004) have identified innovation as a mediator for firm performance, supporting that innovation is a reason for the Hidden Champions' superior performance.

Furthermore, Simon (2009) refers to the development of business processes, with many Hidden Champions stating that innovation includes process innovation that is even more important, *"Expertise lies in the ongoing improvement of processes..."* (Simon 2009, p. 160). Product and process innovation support the long-term orientation of the Hidden Champions as family businesses, with innovations that need to be managed professionally and proactively to raise their full potential. Many family businesses use a reactive approach, though, highlights Wölfer (2010). Din et al. (2013) demonstrated that excellent innovation governance is a confirmed success strategy for Hidden Champions. They succeed through intense cooperation with their customers and achieve superior performance rather than cost leadership. Din et al. (2013) have further identified that high communication intensity and frequency with supply chain partners also have a positive influence on innovation performance. Despite this positive influence on innovation performance, owners or managing partners of Hidden Champions seek independence from their supply chain partners to protect their knowledge base (Din et al. 2013). Dolles (2010) explored the drivers supporting innovation governance in German Hidden Champions. He provided insight into success strategies, among them human resources management (HRM) practices such as recruitment and training programs, leadership styles, organisational structures and corporate networks supporting innovation by ensuring a constant intraorganisational information flow. Noteworthy are the cooperation with local institutions and even co-opetition (Nalebuff and Brandenburger 1997); hence, cooperation with competitors as such, but on an informal level. Both have been identified during interviews with owners or managing partners of Hidden Champions, who confirmed these drivers to mediate innovation, although corporate or interorganisational networks were not considered (Dolles 2010).

The human resources management practices explored above are subject to a further success strategy, the Hidden Champions' high-performance employees for whom they are lauded. Simon (2009) understands training and development of employees as a reason to retain talent and support growth, based on the above-mentioned training programs of employees in the vocational education systems that have a long-standing tradition in the DACH region (Bliem et al. 2016). This distinct approach to employee qualification and the sponsoring of undergraduate academic programmes in dual learning schemes at Universities of Applied Sciences or Universities of Cooperative Education attracts future management trainees, as Venohr et al. (2015) introduced. In Simon's (2020a) most recent analysis of the Hidden Champions' success strategies, the high qualification of the Hidden Champions' employees and continuous innovation have both received stronger emphasis, besides digitalisation requiring a stronger focus (Simon 2020a). An earlier research project by Garaus et al. (2016) supports this analysis. They investigated the influence of human resources management and change management practices such as onboarding, which is the integration of new employees, and training of employees along with team building and network building to understand how Austrian Hidden Champions maintained their long-standing success. Earlier research by Gebauer et al. (2009) emphasised the high relevance of a highly skilled workforce to achieving competitive advantage through service excellence and customer proximity in Swiss Hidden Champions. Research from Voudouris et al. (2000), Witt and Carr (2013) and Schlepphorst et al. (2016) corresponds to the Hidden Champions' managers' understanding of continued training and development, which is vital for the Hidden Champions to maintain and expand the high competence level of their employees. Furthermore, the motivational aspects of employee training providing increased benefits to highly skilled employees were identified and are an essential asset for growth.

Human resources management practices are relevant for the market-leading position of German Hidden Champions, confirms Dolles (2010); the practices he explored are cross-functional teamwork and internal knowledge management that provide a vital communication culture to share ideas and experience to improve the innovation process and therefore maintain and continuously advance their market-leading position.

The third of the identified success strategies is the Hidden Champions' distinct focus on long-term customer relations, that offers inimitable opportunities for business growth and also distinguishes them from their competitors who are focusing on transaction-based business relationships. Venohr et al. (2015), Eckstein and Häfner (2017) as well as Busse and Mayr (2019) state that Hidden Champions are not only manufacturers of specialised products, but they also solve problems for their customers or provide them with full-scale solutions and are known for their high quality aspiration. The Hidden Champions act as the nucleus for innovation and develop customer-specific solutions by proactively participating in the customer's development process. Innovation culture and striving for technological excellence generate a mutual dependency between manufacturer and customer that can create a lock-in situation (Dixit and Nalebuff 1993), preventing customers from switching to competitors, affirm Voudouris et al. (2000) together with Venohr et al. (2015) and Audretsch et al. (2018).

Another success strategy identified in the reviewed literature is globalisation; significant contributors to the globalisation of Hidden Champions are their niche specialisation and their close relation to customers. The opportunities provided by globalisation are apparent in view of the relatively narrow market the Hidden Champions operate in, as it is a necessary condition for their growth, states Witt (2015), confirmed by Audretsch et al. (2018).

The last of the identified success strategies during this literature review is customer proximity, which is also a driver for globalisation, as customers striving for new sources of competitive advantages through globalisation motivate the Hidden Champions to follow in their footsteps (Simon 2009). As a result, foreign production locations can substantially contribute to Hidden Champions' success given improved input factors present Christopher et al. (2011). Scholars such as Merrilees et al. (2001); Venohr and Meyer (2007); Yoon (2013); Witt and Carr (2013); Schlepphorst et al. (2016); Audretsch et al. (2018); Simon (2020) commonly agree that globalisation is imperative for niche-leading businesses to increase their market share. However, they request Hidden Champions to have a strategic framework to master globalisation, understanding the increased risks and complexity of such approaches. The globalisation of Hidden Champions can also be reason for business challenges, such as widespread and complex supply-and-demand networks. How this leads to increased complexity and vulnerability of the Hidden Champions' supply chains will be explained in the following section.

2.3.4 Vulnerability of Global Supply Chains

Globalisation is an undisputed success strategy for Hidden Champions; however, the underlying globalisation of their supply chains often results in higher supply chain complexity and, thus, increased supply chain vulnerability. With tight connections in the Hidden Champions' upstream supply chains, those upstream supply chains are highly vulnerable to disruptive events leading to business disruption or insolvency of the Hidden Champions. Consequently, resulting in forward and downstream disruptions of the Hidden Champions' supply chains (Li et al. 2015). Increased outsourcing of non-core activities in expectation of higher product quality and cost benefits as discussed by Simon (2009) and the offshoring of production to participate in cost advantages as proposed by Jacob and Strube (2008) or Schoenherr et al. (2008) create the tight connections in the upstream supply chains. The resulting dependency on external sources of expertise through single sourcing concerns Chopra and Sodhi (2004) or Wagner and Bode (2006). Along with cultural differences or financial risks identified by Christopher et al. (2011) or Blome and Schoenherr (2011) impacting the Hidden Champions' business performance, the challenges resulting from globalisation lead to increased complexity and vulnerability of the supply chain confirms Tang (2006a, b). Another effect that increases the vulnerability of the supply chains is the forward propagation or diffusion of disruptive events in the supply chains affecting Hidden Champions and the members of their downstream supply chains, which Ivanov et al. (2014) define as the ripple effect. Disruptive events and their consequences toward the vulnerable supply chains are numerous and amplified through the internationalisation of the supply chains, state Steinle and Schiele (2008). However, many of these disruptive events were unpredictable; their impact, however, led to significant economic losses worldwide (Taleb 2007). The Covid-19 pandemic is exemplary for such disruptive events and revealed how the supply chains of the Hidden Champions in the DACH region, initially thought safe, are disrupted because of lockdown measures (Ivanov 2020). Furthermore, to contain the pandemic, entire industries were shut down or discontinued demand in specific industries have disrupted supply chains. The tourism industry, which is an essential customer to many of the Hidden Champions in the cross-border area of the DACH region, stands exemplary for the vulnerability to discontinued demand that has diffused into the Hidden Champions' supply chains resulting in business disruption (Sarkis et al. 2020). To better protect companies against risks such as disruptions in their supply chains, the following section outlines the regulatory frameworks for enterprise-wide risk management including Supply Chain Risk Management.

2.3.5 Regulatory Framework for Supply Chain Risk Management

The previous examples on the vulnerability of the Hidden Champions' global supply chains explain the necessity for risk management and supply chain risk management in particular. With the consequences for the single business and overall economic loss, policymakers in many countries have identified the need for a regulatory framework to identify and manage business risks. This need for an enterprise-wide risk management was motivated by insolvencies of well-known large-sized enterprises e.g. the collapse of German Metallgesellschaft in 1993 (Edwards and Canter 1995) or the Maculan Case, the insolvency of an Austrian construction company in 1996 (Pachl 2016). Based on these needs for a regulatory framework, legal requirements have been established demanding an enterprise-wide risk management approach to avoid business discontinuity.

Enterprise risk management (ERM) is a comprehensive approach to identify and manage corporate risks and the overarching concept to Supply Chain Risk Management state scholars like Blome and Henke (2009); Sodhi and Tang (2011); Blome and Schoenherr (2011); Sodhi et al. (2012); Curkovic et al. (2013); and Gredal et al. (2016). Table 2.1 shows the history of regulatory frameworks identified in the DACH region requiring businesses to establish an enterprise-wide risk management. The intention for the development of ERM was based on the purpose of managing financial risks. The monetary risks in the supply chain, however, are to be managed and mitigated by Supply Chain Risk Management which is supported by scholars such as Rao and Goldsby (2009); Sodhi and Tang (2011); Wagner and Bode (2008) and Heckmann et al. (2015). The focus of Supply Chain Risk Management as portrayed above is the approach to be followed in this study and is discussed further in the next section.

Table 2.1 Legal Requirements for Enterprise Risk Management

Year of publication	Country	Description	Aim of the regulation
1997	**Austria**	**IRÄG** Insolvenzrechtsänderungsgesetz (Insolvency Amendment Act)	Considers significant changes in financial reporting and legally requires risk management.
1998	**Germany**	**KonTraG** Gesetz zur Kontrolle und Transparenz im Unternehmensbereich (Corporate Control and Transparency Act)	Supervisory boards are required to execute effective monitoring of the enterprise risk management process and to intensify the legal liability of management boards.
2007	**Switzerland**	**Obligationsrecht,** Durchführung Risikobeurteilung (Art. 663b Ziff. 12 OR) (The Swiss Code of Obligations Performance of a risk assessment) (Art. 663b, Para. 12 SCO)	Requires the existence of an internal control system and comprehensive reporting by the auditor to the Board of Directors.

2.4 Supply Chain Risk Management and Hidden Champions

Since the beginning of the new millennium, report Zsidisin and Henke (2019), Supply Chain Risk Management has gained increasing attention from both academics and practitioners. Numerous disruptive events influencing the economic performance of businesses worldwide have triggered this attention that is boosted by the Covid-19 pandemic driven supply chain disruptions. The following sections discuss the resilience of Hidden Champions in general and present different definitions of Supply Chain Risk Management along with different scopes of supply chains and different core elements of different Supply Chain Risk Management processes identified during this literature review. Due to the lack of a universal definition for Supply Chain Risk Management, which is to be seen in the context of the unclear scope of the supply chain and the varying elements of the Supply Chain Risk Management process, this section presents a definition for Supply Chain Risk Management that is used in the context of this thesis. Hidden Champions are commonly seen as being highly resilient against business disruptions, however not immune as noticed by Simon (2009); therefore, the next section will give insight into their resilience and how prepared they are to manage and mitigate disruptive events in their supply chains.

2.4.1 Resilience of Hidden Champions

Christopher et al. (2003) define resilience as the ability inherent to a system to return to its initial or expected state after a disruptive event. Simon (2017) rates the German Hidden Champions as being highly resilient estimating that nearly all of them survived the economic recession following the 2007–2008 financial crisis. Their high resilience and contribution to economic development both in GDP generation and employment make the Hidden Champions an attractive model to follow for the implementation in emerging markets, where the emergence of small and medium-sized businesses should receive institutional support declare Landau et al. (2016). Reasons for this high resilience identified in the reviewed literature are seen in the Hidden Champions' distinct risk aversion and their financial health argue Walravens and Filipović (2013), supported by Lee and Wang (2017), noting that financial health is of paramount importance. Neubauer (2011) disagrees with the above laudation on the Hidden Champions' resilience and identifies that smaller Hidden Champions have difficulties in responding promptly to disruptive events because their organisations are not prepared. Also, Simon asserts that it must be understood that even *"... Hidden Champions are not miracle companies. They are not immune to crises..."* (Simon 2009, p. 24). Many of the findings above lauded the resilience of Hidden Champions against disruptive events. Before understanding how Supply Chain Risk Management helps to increase the resilience of a supply chain, a workable definition of Supply Chain Risk Management applicable in Hidden Champions is needed, which are discussed in the following section.

2.4.2 Ambiguity of Supply Chain Risk Management Definitions

The research on Supply Chain Risk Management has significantly developed during the last 20 years (Zsidisin and Henke 2019). However, *"a well-grounded, unified and universally recognised Supply Chain Risk Management definition is presently missing"* state Kilubi and Haasis (2015, p. 36). The overview in Table 2.2 displays the most cited Supply Chain Risk Management definitions identified.

Table 2.2 Most cited Supply Chain Risk Management Definitions

Author	# of Citations	SCRM definition
Jüttner et al. 2003, p. 201	1,836	"the identification and management of risks for the supply chain, through a coordinated approach amongst supply chain members, to reduce supply chain vulnerability as a whole."
Norrman and Jansson 2004, p. 436	1,335	"Supply chain risk management is to [collaborate] with partners in a supply chain risk management process tools [sic] to deal with risks and uncertainties caused by, or impacting on, logistics related activities or resources".
Manuj and Mentzer 2008, p. 205	1,054	"Global supply chain risk management which is the identification and evaluation of risks and consequent losses in the global supply chain, and implementation of appropriate strategies through a coordinated approach among supply chain members with the objective of reducing one or more of the following – losses, probability, speed of event, speed of losses, the time for detection of the events, frequency, or exposure – for supply chain outcomes that in turn lead to close matching of actual cost savings and profitability with those desired".

The above statement of Kilubi and Haasis (2015) is supported by other Supply Chain Risk Management researchers (Manuj and Metzner 2008; Rao and Goldsby 2009; Diehl and Spinler 2013; König and Spinler 2016; Baryannis et al. 2019). Further research asserts that the missing unified definition hinders the implementation of Supply Chain Risk Management in businesses (Diehl and Spinler 2013). This definitional gap makes it challenging to create a shared understanding among researchers and practitioners (Sodhi et al. 2012; Diehl and Spinler 2013; Ho et al. 2015) that could foster the implementation of Supply Chain Risk Management. This various Supply Chain Risk Management definitions indicate that, instead of giving a universally accepted definition for Supply Chain Risk Management, the scope of the underlying supply chain and the key elements of the Supply Chain Risk Management process steps are inconsistent within these definitions.

2.4.3 Differing Scope of the Underlying Supply Chain

Different Supply Chain Risk Management scholars define the scope of the underlying supply chain in a way that is too extensive and therefore too complex for a realistic study. Defining the scope of the supply chain as part of this discussion will help to generate a mutual understanding of the area researched, the focus on the entire supply chain would be too extensive and therefore too complex also supported by Ziegenbein (2007). For example, the scope of the supply chain used in the research of Jüttner et al. (2003) ranges among supply chain members, whereas the identification of all the supply chain participants involved appears to be unrealistic. Other researchers discuss a global supply chain, which most supply chains certainly are today; a specific depth or width of the supply chain is not further discussed (Manuj and Metzner 2008). Finally, Norrman and Jansson consider a more quantifiable, a dyadic business relation, or *"preferably, recommend a supply chain of three or more companies."* (2004, p. 436). This discussion can be concluded with the findings of Pfohl et al. (2010) understanding that implementing a Supply Chain Risk Management process within an entire supply chain is virtually impossible due to its widely interwoven network structure which would require the central company, governing the supply chain, to know about its entire supply chain, its structure and its supply chain partners. Implementation within the upstream supply chain, ranging from the Hidden Champions, towards their first and second tier suppliers, and the downstream supply chain, from the Hidden Champions towards their customers, however, is realistic.

2.4.4 Differing Schools of Supply Chain Risk Management Processes

A unified definition of Supply Chain Risk Management is still not existing. Following Manuj and Mentzer (2008), a clear understanding based on improved communication between researchers and practitioners is needed, particularly within the specific context of Hidden Champions, to foster the implementation of Supply Chain Risk Management. Christopher et al. (2003) define Supply Chain Risk Management processes as a two-step approach consisting of identifying risks and implementing appropriate risk management strategies. What is defined in the following as the Christopher school (Table 2.3), is widely supported by other researchers (Colicchia and Strozzi 2012; Golgeci and Ponomarov 2013; Heckmann et al. 2015, Hohenstein et al. 2014).

Table 2.3 Key Elements of the Christopher-School

Article	Year	Key Elements of SCRM				Scope of the SC	Approach
Christopher et al.	2003	Describe your SC	Vulnerability Self-Assessment	Evaluating the Implications	Identifying Actions	End-to-end-SC	x
Colicchia and Strozzi	2012	Identification	x	x	Implement appropriate Strategies	Among SC members	x
Golgeci and Ponomarov	2013	Identification	x	x	Implement appropriate Strategies	Among SC members	x
Hohenstein et al.	2014	Identification	x	x	Manage risks in global a SC	Global SC	proactive
Heckmann et al.	2015	Identification	Assessment	x	x	x	x

This stands in contrast to researchers like Norrman and Jansson (2004) or Kleindorfer and Saad (2005) who identify a Supply Chain Risk Management process to be organised in three steps: risk identification or specification, risk assessment and risk mitigation. An overview of the Norrman-School to Supply Chain Risk Management is shown in Table 2.4.

Table 2.4 Key Elements of the Norrman-School

Article	Year	Key Elements of SCRM				Scope of the SC	Approach
Kleindorfer and Saad	2005	Specifications	Assessment	x	Mitigation	x	x
Norrman and Jansson	2004	Identification	Assessment	Management	x	Collaborate with partner in a SC	proactive

An iterative process to control or monitor the results achieved and to adapt on a by case basis leads to continuous improvement strengthening the Supply Chain Risk Management process which is supported by scholars like Pfohl et al. (2010); Tummalla and Schoenherr (2011); Curkovic et al. (2013); Diehl and Spinler (2013); Cagnin et al. (2016); and Qazi et al (2018). An overview of the Pfohl-School to Supply Chain Risk Management is presented in Table 2.5.

Table 2.5 Key Elements of the Pfohl-School

Article	Year	Key Elements of SCRM				Scope of the SC	Approach
Cagnin et al.	2016	Identification	Assessment	Evaluation	Mitigation	Among SC members	proactive
Curkovic et al.	2013b	Identification	Analysis	Evaluation	Treatment	Collaborate or coordinate with supply partners	x
Diehl and Spinler	2013	Identification	Assessment	Mitigation	Controlling	Upstream and downstream SC partners	x
Pfohl et al.	2010	Identification	Assessment	Mitigation	Controlling	All partners and relationships	proactive
Qazi et al.	2018	Identification	Analysis and Evaluation	Treatment	Re-Assessment	Among SC members	x
Tummala and Schönherr	2011	Identification	Measurement & Assessment	Evaluation	Control and Monitoring	Upstream supplier - focal firm - downstream suppliers	x

Following these threads to an amplified and iterative Supply Chain Risk Management process motivates to further investigate in the research of Tummala and Schoenherr (2011); Curkovic et al. (2013); Cagnin et al. (2016); Trkman et al. (2016). These articles present a Supply Chain Risk Management definition given by the International Organization for Standardization (ISO) based on the ISO 31000 (ISO 2018) standard. Scholars like Ataseven et al. (2014) and de Oliveira et al. (2017) discuss the principles and generic guidelines on risk management provided by ISO 31000 (ISO 2018) that could find growing attention as this standard risk management procedure is part of the ISO 9001:2015 revision of the quality management standard (ISO 2015). This quality management standard has been adopted in earlier revisions by many Hidden Champions in the mechanical engineering industry in the DACH region.

2.4.5 Definition of Supply Chain Risk Management

The missing unified definition of Supply Chain Risk Management, the lack of understanding among researchers and practitioners (Sodhi et al. 2012; Diehl and Spinler 2013; Ho et al. 2015) and thus its impeded implementation in companies (Diehl and Spinler 2013) are strongly associated with the unclear scope of the underlying supply chain discussed by Otto and Kotzab (2001); Jüttner et al. (2003); Zsidisin (2003) and Wagner and Bode (2006). The Supply Chain Risk Management definitions found in the literature (Jüttner et al. 2003; Norrman and Jansson 2004; Manuj and Mentzer 2008; Zsidisin and Ritchie 2009) are visualised in the form of a word cloud which articulates the many different elements identified, which manifests the earlier discussed ambiguity of Supply Chain Risk Management definitions (Fig. 2.5).

Fig. 2.5 Visualisation of Supply Chain Risk Management Definitions

Based on the large number of different contents in the visualisation of the various definitions and the requirements identified from the semi-structured interviews, the author defines cooperative Supply Chain Risk Management as follows:

> *"Supply Chain Risk Management is a proactive and iterative process that is based on interorganisational cooperation between the company in focus, its first and second-tier suppliers, and its internal processes up to delivery to the customers. The risks to be managed within this supply chain are the uncertainties for the company in focus. They need to be identified, analysed, evaluated and managed through appropriate strategies that are adapted to the business needs of the entire company. This interorganisational Supply Chain Risk Management process is continuously controlled and monitored to guarantee effective risk management strategies in an unstable business environment."*

This definition will be followed during the further course of this research project. Given this definition of Supply Chain Risk Management covers the depth and width of the Supply Chain Risk Management process, the following section will focus on the Supply Chain Risk Management literature relevant to Hidden Champions.

2.4.6 Supply Chain Risk Management Research in Hidden Champions

This section will discuss the different streams in Supply Chain Risk Management research and the relevance for researching it in the field of Hidden Champions. Existing research on risk management in Hidden Champions from Lebedev (2013) and Mietzner et al. (2018) identifies financial risk management that, according to Huber and Gith (2009), improves a company's credit rating. Michaels and Grüning (2017) and Ulrich (2011) identify corporate social responsibility and corporate governance. Pundy (2015) or Schwass and Glemser (2016) identified organisational culture and reputational risk in Hidden Champions. Finally, Zhexembayeva (2013) discusses risk mitigation in sustainable Hidden Champions in central and eastern Europe (CEE), where risk mitigation is seen to neither create value nor bottom-line growth. To avoid business discontinuity due to increasing sustainability-related disruptions, the Hidden Champions in central and eastern Europe must effectively manage and mitigate these sustainability-related risks by implementing rigid risk management procedures. While no literature on the phenomenon of Supply Chain Risk Management within Hidden Champions in the DACH region was available, the author extended this research to small and medium-sized enterprises and family businesses since these found entrance in both Supply Chain Risk Management and Hidden Champions research over the last years which the next section demonstrates.

2.4.7 Small and Medium-Sized Enterprise Perspective

Given their relevance to the global economy, small and medium-sized enterprises have been extensively researched. The shortages of resources along with the professionalism of processes and their organisational culture have been the focus in earlier studies relating to Supply Chain Risk Management. Finch (2004) and Villena and Gioia (2018) identified the increasing risk exposure of large-sized enterprises when they are in a business relation with small and medium-sized enterprises. However, according to Faisal et al. (2007), the implementation of Supply Chain Risk Management processes in small and medium-sized enterprises is subject to many barriers. On part of the small and medium-sized enterprises, distrust and "*information distortion*" (Faisal et al. 2007, p 594) is hindering information sharing enabled by introducing Supply Chain Risk Management processes and creating overall benefits along the supply chain. The terminological ambiguity of Supply Chain Risk Management combined with limited experience and training in small and medium-sized enterprises give focus to only a limited number of risk management activities, such as local sourcing; safety stocks and financial reserves which is supported by Ellegard (2008) and Rao and Goldsby (2009). Proactive risk management practices for small and medium-sized enterprises were explored by Chiarini (2017), along with them the ISO 9001:2015 standard (ISO 2015). In this research, the ownership of the risk management process remained unclear while Chiarini (2017) surveyed quality managers identifying significant risk sources in small and medium-sized enterprises. The above articles of Finch (2004); Faisal et al. (2007); Ellegard (2008); Rao and Goldsby (2009); Chiarini (2017) or Villena and Gioia (2018) highlight the research on Supply Chain Risk Management activities in small and medium-sized enterprises; they also provide further insights into the barriers to implementing Supply Chain Risk Management in small and medium-sized enterprises, the ambiguity of Supply Chain Risk Management definitions and the unclear ownership of Supply Chain Risk Management processes in small and medium-sized enterprises. This section concludes with the findings of Verbano and Venturini (2013), stating that risk management and, in particular, Supply Chain Risk Management in small and medium-sized enterprises has only gained interest since the beginning of the new millennium and are still not sufficiently studied. The next section will provide understanding of how family businesses manage and mitigate supply chain risks.

2.4.8 Family Business Perspective

Research on family businesses has enjoyed considerable attention and has demonstrated the importance of this field of research. It has further helped to understand the family businesses' distinct behaviour. Similar to the previous section on small and medium-sized enterprises, this section reviews and discusses the phenomenon of Supply Chain Risk Management in family business literature. In many family businesses, the owners themselves manage the supply chain in a dual role besides their task as owner-managers assert Ritchie and Brindley (2000) or Falkner and Hiebl (2015), giving them undoubtedly an elevated level of awareness regarding supply chain risks. McNamara has identified the owner-managers dominance in the organisation as *"founder's syndrome"* (1998, p. 38), which, according to Sutherland and Purg (2013), often leads to a reactive rather than a proactive strategy-driven approach to Supply Chain Risk Management. The research of Ritchie and Brindley (2000) and Ellegard (2008) supports this by identifying Supply Chain Risk Management practices to be informal or missing altogether, and by Finch (2004), who notes that the skills shortage of owner-managers is apparent and has myopic effects on the requirements for business process excellence. The findings of Jayaram et al. (2014) also support this by pointing towards higher supply chain performance in non-owner-managed family businesses compared to owner-managed family businesses. Jayaram et al. (2014) continue to discuss the performance of supply chain management (SCM) as a system, which is supported by establishing trustful communication between the family businesses and their supply chain partners. The underlying theoretical perspectives influencing the above findings are discussed in the next section.

2.4.9 Theoretical Perspectives

This review has identified different theoretical perspectives. The majority of studies focus on innovation in Hidden Champions; hence, theoretical backgrounds to innovation are discussed. Dolles (2010) and Din et al. (2013) discuss the controversial theory of Schumpeter (1942) identifying firm size as an important factor in innovation performance. In studies focusing on success factors, several theoretical perspectives are found. Gruenwald (2013) investigates in success factors that support the growth of Hidden Champions. Voudouris et al. (2000) are positioned in the resource-based view (RBV) of the firm (Rumelt 1984), identifying unique resources and capabilities that enable success. A similar position can be found in the study of Garaus et al. (2016), focusing on human resources management systems that base their studies on the resource-based view (Barney 1991) and identify a long-term focus for managing human resources to achieve a sustained advantage. The resource-based view seeks to explain how businesses can achieve and uphold competitive advantage and is most prevalent in strategic management research. The fundamental concept of the resource-based view is the combination of resources; however, being valuable, rare, inimitable and non-substitutable (VRIN) to achieve sustained competitive advantage (Barney 1991). Hence, the resource-based view could be a possible theoretical framework for researchers to demonstrate Hidden Champions being successful in their market based on products and capabilities that their competitors cannot imitate. Other researchers, like Pundy (2015), focus on cultural similarity in the DACH region or base their study on cultural value orientation developed by Schwartz (2006). Finally, studies on family businesses describe those as risk-averse from an entrepreneurial standpoint to safeguard the owning family's status in business and society, which is based on the socioemotional wealth (SEW) theory (Berrone et al. 2012; Gómez-Mejía et al. 2007), that, according to Maloni et al. (2017), lead to less vulnerable supply chain strategies that avoid outsourcing and global sourcing. Key researchers to socioemotional wealth research (Gomez-Mejia et al. 2007; Berrone et al. 2012) indicate that socioemotional wealth positively stimulates family businesses' performance on the level of management processes, business strategies, corporate governance, stakeholder relationships, and business venturing. However (Martínez-Romero and Rojo- Ramírez 2016) question the positive impact on supply chain performance when applying the socioemotional wealth concept. While Hidden Champions enjoy greater autonomy from institutional investors due to their financial health identified by Walravens and Filipović (2013) or Lee and Wang (2017), this results in more risk-taking strategies in family businesses to preserve the socioemotional wealth as discussed by Gómez-Mejía et al. (2007). Based on these theoretical perspectives, this review concludes that the topic area is largely understudied, bearing significant insufficiencies in research, both in thematic context and in theoretical perspective, hence requiring further study. The gaps identified in this literature review will be discussed in the next section.

2.5 Gaps in the Literature

With both research areas, Hidden Champions and Supply Chain Risk Management, being in their adolescence, studies on Hidden Champions mostly focus on success strategies defined by Simon (2009), whereas Supply Chain Risk Management researchers like Zsidisin and Henke (2019) discuss definitions and necessary steps in the Supply Chain Risk Management process itself. In this section, the gaps identified in the literature are discussed to define areas for further research. Dolles (2010) and Din et al. (2013) identify innovation as the most prominent of the Hidden Champions' success strategies, both product and process innovation, in addition Wölfer (2010) discusses proactive cooperation or interorganisational networks having the most positive influence on innovation performance. A similar approach can be found at Voudouris et al. (2000) and Garaus et al. (2016) in their research of the high skill level and motivation among the Hidden Champions' employees, a success strategy based on continuous training and development. Within training and development, cross-functional teamwork and network building enable a vital communication culture, across organisational boundaries, that helps to share ideas and experiences to continuously improve products and processes. According to Andreeva et al. (2016), expanding the communication culture across organisational boundaries mediates proactive participation in the customer development process. The benefits of sharing information across organisational boundaries, including understanding customer targets (Wagner 2010), are part of the close relation to customers, which is also among the success strategies of Hidden Champions. These influences on the Hidden Champions' success strategies mutually reinforce each of the individual success strategies. A vital communication culture, proactive interorganisational cooperation as well as cooperation towards customers, and finally intraorganisational networks support active internal cooperation and fruitful cooperation with the customers. However, cooperation and communication with supply chain partners to manage and mitigate disruptive events in the upstream supply chain is not yet a priority among the Hidden Champions. Globalisation as the last of the identified success strategies, which is imperative for niche-leading businesses to increase their market share (Simon 2017), leads to increased complexity and thus vulnerability of the supply chain as confirmed by Christopher and Lee (2004); Jüttner (2005) and Steinle and Schiele (2008). Disruptive events in such complex supply chains are pervasive and have amplified over time. While Hidden Champions often cooperate with smaller partners, who, according to Ellegard (2008), suffer from a lack of resources and experience, the upstream supply chain seems to be increasingly vulnerable (Finch 2004) to disruptive events; further research by Villena and Gioia (2018) states that firms lack knowledge about who their upstream supply chain partners are. Hence, Din et al. 2013 have identified that Hidden Champions require a strategic layout for globalisation, including identifying and empowering of suitable partners for supply chain cooperation and risk management at a global level.

In summary, based on the identified Hidden Champions success strategies, close cooperation and interorganisational networks lead to a better understanding of the relevance of Supply Chain Risk Management for the success of the Hidden Champions and their upstream supply chain partners. Improved communication within an interorganisational network can lead to immediate actions in case of disruptive events by initiating countermeasures among the supply chain members. Based on the literature gaps previously identified, the following section will identify areas for further research.

2.6 Identifying Areas for further Research

Business without risk is not possible. However, management of these risks is essential to eliminate the negative impacts of risks, states Faisal et al. (2007). This research supports the findings of Faisal et al. (2007), stating that there is a need to introduce Supply Chain Risk Management processes in Hidden Champions, given their economic and social importance worldwide. This section addresses the gaps in the literature. It proposes two areas for further research concerning the strategic layout of Supply Chain Risk Management across organisational borders and the generation of future knowledge based on previous experiences that are essential to support the management and mitigation of disruptive events in the Hidden Champions' upstream supply chains.

2.6.1 Supply Chain Risk Management Cooperation

Close customer relations enable the Hidden Champions to develop innovative products providing unique technical solutions to their customer, which is emphasised by Venohr et al. (2015). Dolles (2010) identified the Hidden Champions' vital communication culture that expands into the customers' organisation allowing for the proactive participation in the customers' development process. This strong interdependency within the supply chain through cooperation discussed by Xu and Beamon (2006) leads to the alignment of the Hidden Champions targets with those of their customers, confirmed by Wagner (2010) or Bals and Turkulainen (2021). Interorganisational networks have been confirmed to enrich innovation by communicating intensively and sharing information among network participants in the upstream supply chain. Yet, Dolles' (2010) empirical findings see corporate or interorganisational networks as unimportant. The question here is whether this is rooted in the Hidden Champions' distrust according to Faisal et al. (2007) and protectionism against exploitation of innovation raised by Din et al. (2013) or a lack of importance. The benefits of sharing information across organisational boundaries including the understanding of customer targets, however, are identified as being success strategies of Hidden Champions. Research by Friday et al. (2018) on cooperation in supply chains emphasises alignment of strategy to limit the damage to the disrupted supply chain. This is supported by Brüning (2019) stating that establishing trust among supply chain partners to achieve mutual success that lies beyond what can be achieved by the Hidden Champions alone. Therefore, enhanced communication and transparency on the target alignment between Hidden Champions and their supply chain partners appear to be vital for the success of the Supply Chain Risk Management approach of the entire supply chain. Once supply chain members understand the relevance of improved communication for the success of Supply Chain Risk Management in the upstream supply chain, Wagner (2010) asserts that this can motivate them to take immediate action in case of disruptive events and initiate countermeasures. Establishing trust and aligning the Supply Chain Risk Management strategy among Hidden Champions and their supply chain partners have not been topic of Supply Chain Risk Management research. Hence, this research project seeks to understand the cooperation between the Hidden Champions and their supply chain partners to manage and mitigate disruptive events in the upstream supply chain which requires intensive communication and instant information sharing.

2.6.2 Supply Chain Risk Management Training

According to Ellegard (2008), the smaller partners the Hidden Champions often cooperate with, often suffer from a lack of resources and experience. Hence, the upstream supply chain seems to be increasingly vulnerable to disruptive events due to the skills shortage of family business owner-managers that Finch (2004) identified. This increased vulnerability threatens the entire supply chain of the Hidden Champions. A strategic layout including the empowerment of supply chain partners through training and development on a global level is required to master this vulnerability. This strategic layout for generating future knowledge among employees is identified as of strategic importance. Shared learning in the form of training and development across organisational boundaries requested by Bessant et al. (2003) is a key component requiring trust and intense cooperation among supply chain members, also confirmed by Spekman et al. (2002). Moreover, this study seeks further understanding as to whether employees in Hidden Champions receive the same training and development to master disruptive events in their supply chain as they do to contribute to innovation according to Dolles 2010; Din et al. 2013). It is further unclear from the previous review how Hidden Champions empower their supply chain partners, which Mohr and Sengupta (2002) requested and whether the existence of a Supply Chain Risk Management process at these supply chain partners is mediating the management and mitigation of disruptive events throughout the entire supply chain. With this section identifying the areas for further research to address the literature gaps, the next section will provide the research question and constituting sub-research questions of this research project.

2.7 Research Question

The findings of the literature review in the previous sections provide the emergence of a research question that is reported upon in this thesis:

> **How does interorganisational cooperation between Hidden Champions and their supply chain partners mediate the management and mitigation of disruptions occurring in the upstream supply chain of Hidden Champions in the DACH region?**

To fully answer this research question, it will be necessary to address the following two constituent sub-research questions:

i. How are the Hidden Champions cooperating with their supply chain partners to manage risks in the Hidden Champions' upstream supply chain?
ii. How are the Hidden Champions' employees and the supply chain partners trained and developed to understand and apply the Supply Chain Risk Management processes available?

This research question will seek an understanding of the phenomenon of Supply Chain Risk Management cooperation between Hidden Champions and their supply chain partners. Answering this research question will explain whether interorganisational cooperation, which requires training and development of the Hidden Champions' employees and their supply chain partners, will mediate an efficient Supply Chain Risk Management process. In case of a supply chain disruption, efficient management and mitigation of the event require understanding the Hidden Champions' Supply Chain Risk Management processes and swift and seamless communication to ensure business continuity.

2.8 Chapter Summary

This chapter, Chapter 2, presented an initial review of the extant literature on Hidden Champions and Supply Chain Risk Management. It thus displayed success strategies, such as high-performing employees and close cooperation with the customers as the fundament for their success and their high significance for the economy. A further success strategy is their globalisation though increasing the vulnerability of their upstream supply chain to disruptive events. Given this vulnerability, the author reviewed the literature on Hidden Champions' approaches to manage and mitigate these disruptive events. The findings display a lack of formalised Supply Chain Risk Management processes and a missing theoretical foundation for Supply Chain Risk Management. This review concluded by defining this research's question that demands further research on the need for interorganisational training and development of the Hidden Champions and their supply chain partners' employees to foster communication and cooperation to improve the interorganisational management and mitigation of disruptive events.

In the next chapter, the theoretical framework for this research project is defined. The previously defined research question and its constituent sub-research questions have a significant influence on the design of this theoretical framework to help to understand and explain how the Hidden Champions that participate in this research project apply and disseminate the Supply Chain Risk Management processes in their organisations and how the supply chain partners participate in these processes. The later development of a conceptual framework is supported through an appropriate theoretical framework.

3 Theoretical Framework of this Research

The previous chapter portrayed the success strategies of Hidden Champions, based on their ability to rise above their competitors through innovative processes, innovation governance (Dolles 2010), and close customer proximity, making them essential, often indispensable, business partners for their customers (Simon 2009). Customer proximity induces Hidden Champions to follow the globalisation approaches of their customers, resulting in dependence on external partners in the supply chain that increases the complexity thus vulnerability of their supply chains. To master this supply chain complexity and thus reduce the vulnerability against disruptive events, Hidden Champions require a strategic framework to master the challenges of globalisation, as recommended by Din et al. (2013) and Simon (2020), including Supply Chain Risk Management.

This chapter describes the theoretical tools that helped to frame and guide this research project. The focus is to understand and explain how the Hidden Champions participating in this research project apply and disseminate the Supply Chain Risk Management processes in their organisations and how the supply chain partners participate in these processes. The author addresses these questions by adopting the supply chain practice view (SCPV), a newly developed approach embracing several theories used in supply chain management research proposed by Carter et al. (2017). The supply chain practice view supports the considerations of this research project in two ways: firstly, it compensates for the shortcomings of theories commonly used in supply chain management research, such as the resource-based view (Barney 1991), by spanning across organisational boundaries (Carter et al. 2017). Second, the novelty of the supply chain practice view offers

© The Author(s), under exclusive license to Springer Fachmedien Wiesbaden GmbH, part of Springer Nature 2026

M. Mayer, *Spanning Organisational and Geographical Boundaries*,
https://doi.org/10.1007/978-3-658-50453-3_3

the opportunity to understand and explain significant problems in supply chain management research from a perspective that includes the actors contributing to the supply chain (Pagell et al. 2022), thus responding to the call of Sodhi et al. (2012) for more empirical research on Supply Chain Risk Management.

3.1 Prevailing Theories in the Reviewed Literature

Theoretical frameworks recapitulate the prevailing theories in the relevant research area and provide a basis for data analysis and interpretation of the results. They summarise the concepts and theories published earlier on which the background of this study is built (Kivunja 2018). The theoretical framework justifies the research effort based on the literature review that identifies the lack of research on Supply Chain Risk Management in Hidden Champions and Supply Chain Risk Management spanning organisational boundaries in particular. The literature review further identified the prominent theories in Hidden Champion and Supply Chain Risk Management research, identifying mainly the success strategies compared to their competitors as a blueprint to follow. The resource-based view as a prevailing theory in strategic management research seeks to explain how businesses such as the Hidden Champions can achieve and uphold competitive advantage. Therefore, the resource-based view could be a possible theoretical lens for this research as it suggests that purchasing and supply chain management offer characteristics that lead to a sustained competitive advantage for the company in focus (Barney 2012). While the supply chain perspective is at least dyadic, within this research, it is not sufficient to only focus on resources being valuable, rare, inimitable and non-substitutable, hence focusing on the individual organisation itself. The term supply chain, a complex network of different organisations, implies taking an interorganisational perspective. Therefore, this study applies an interorganisational perspective to Supply Chain Risk Management. This perspective spanning the organisational boundaries of the Hidden Champions and their supply chain partners, is one factor not considered in the resource-based view. Allred et al. (2011) support this interorganisational perspective by showing that cooperation among the supply chain partners is required to achieve superior performance. Further theories identified discussed the risk-taking behaviour of the owning family and its stimulus on the business performance (Gómez-Mejía et al.

2007; Berrone et al. 2012). However, the author questions the positive impact on performance of the supply chain influenced by socioemotional wealth. While preserving socioemotional wealth for the next generation is achieved through power and influence staying in family hands (Gomez-Mejia et al. 2011), it only concentrates on the family business itself, which will negatively affect interorganisational cooperation. Supply Chain Risk Management focusing on dyadic business relations, or a network-oriented focus is not represented nor sufficiently discussed in most of the established theoretical considerations, asserts Spina et al. (2016). Moreover, it is precisely the interorganisational relationships in customer-supplier relations that lead to increased resilience of the supply chain to disruptive events, confirm Kaufmann et al. (2018). For this research that seeks to understand Supply Chain Risk Management spanning the organisational boundaries of Hidden Champions, an adequate theoretical perspective needs to support the interorganisational approach of this study. Hence, this study discusses a theoretical lens in the following section, suitable for understanding and explaining new theoretical approaches in supply chain management supporting the interorganisational perspective of this study found in the supply chain practice view as introduced by Carter et al. (2017).

3.2 Supply Chain Practice View: a Theoretical Framework

The supply chain practice view contrasts with the resource-based view that, based on strong isolating mechanisms, limits itself to using only firm-oriented resources and therefore lacks benefit from the advantages arising from cooperation (Fig. 3.1).

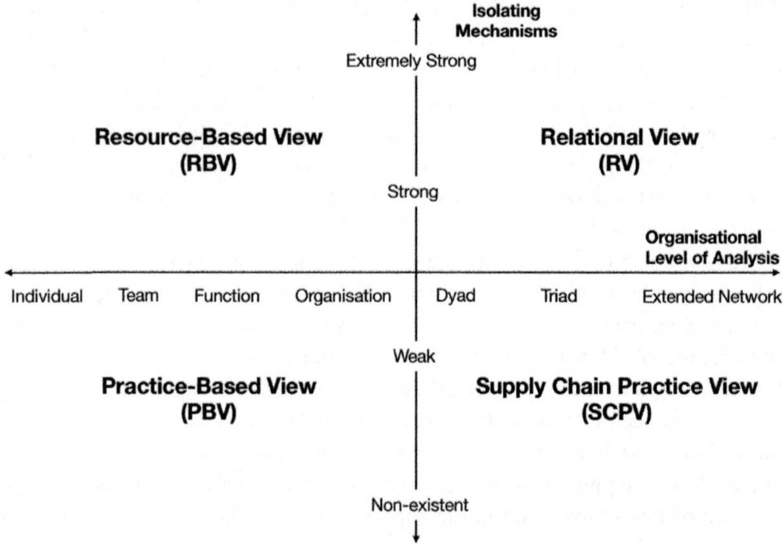

Fig. 3.1 Theories influencing the Supply Chain Practice View. (Adapted from Carter et al. 2017, p. 118)

The supply chain practice view builds on the practice-based view proposed by Bromiley and Rau (2014). In their later publication (Bromiley and Rau 2015), they evaluate the usefulness of the resource-based view in supply chain management and justify improvements in process performance using imitable and transferable practices resulting from a broad organisational level of analysis as a result of cooperation within dyadic or network-oriented business relations. Shaw et al. (2021) discuss the supply chain practice view considering the relational view. The relational view was offered by Dyer and Singh (1998), they understand that resources can span organisational boundaries even though critical to an organisation. This understanding motivates Shaw et al. (2021) to promote cooperative approaches in the relationship between the supply chain members, hence applying supply chain management practices that span the organisational boundaries of the firm. Resulting from the different theoretical influences, the supply chain practice view focuses on increasing the performance of the interorganisational supply chain, including that of Supply Chain Risk Management. To understand the concept of performance, supply chain practice view scholars (Carter et al. 2017; Kosmol et al. 2019; Shaw et al. 2021) distinguish between economic performance, including operational, financial, and market-related performance and social performance in addition to ecological performance. The first three forms of performance play a significant role for the Hidden Champions, primarily organised as family-owned businesses. Thus, within the framework of economic performance, the financial and the market-related performance are of immense importance for Hidden Champions to maintain their niche-leading position and to guarantee the business's long-term success. In addition, the close relationship with the customer is of decisive importance for the success of the Hidden Champions, confirmed by Felden et al. (2019) and Schlüter (2018). As a result, customer satisfaction plays a significant role, expressed in the operational performance of the Hidden Champions, i.e., the ability to deliver on time and in full quantity (OTIF) to the customer, that McLean (2017) sees as the ultimate objective to achieve in a supply chain. The supply chain practice view aims to promote cooperation in managing the supply chain across organisational boundaries (Carter et al. 2017). Thus, applying it to Supply Chain Risk Management leads to a more potent Supply Chain Risk Management across the entire supply chain rather than only for the Hidden Champion leading the supply chain (Pagell et al. 2022). The increasingly strong dependence on external partners in the Hidden Champions' upstream supply chain requires a theoretical lens that spans the organisational boundaries of the supply chain members rather than focusing on inimitable capabilities leading to sustained competitive advantage (Barney 1991). Applying the supply chain practice view as a theoretical lens magnifies cooperative relationships increasing supply chain resilience. Applying an interorganisational approach involving supply chain partners in the Supply Chain Risk Management process helps identify disruptions in the upstream supply chain in good time and thus manage and mitigate them more effectively. This mediation of Supply Chain Risk Management and thus increasing the supply chain resilience of the Hidden Champions and their supply chain partners will positively affect the supply chain performance.

3.3 Interorganisational Practices

To position the Supply Chain Risk Management practices identified in this study, the author adopts a continuum presented by Carter et al. (2017) consisting of two dimensions (Fig. 3.2).

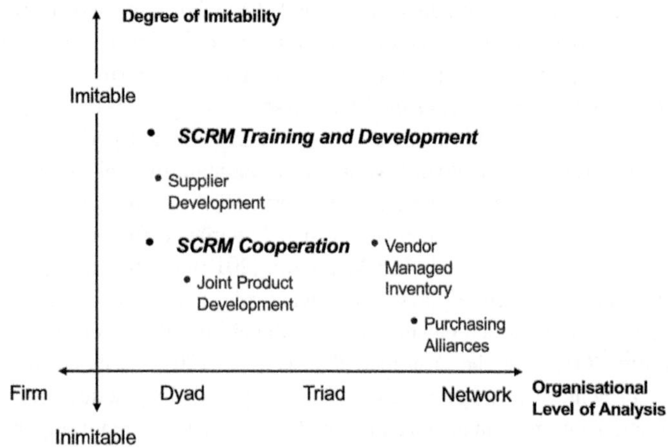

Fig. 3.2 Framework of this Research. (Adapted from Carter et al. 2017, p. 118)

The first dimension of this continuum describes the organisational level on the horizontal axis, which spans the organisational boundaries of the Hidden Champions to the dyad or the network. The second dimension on the vertical axis represents the degree of imitability or transferability of the used Supply Chain Risk Management practices. Supply Chain Risk Management training and development of the Hidden Champions' employees and those of their supply chain partners that require interorganisational cooperation are imitable and transferable practices that span organisational boundaries, similar to activities such as supplier development (Helmold 2014) or the joint product development discussed by Din et al. (2013) or Parast et al. (2019). In the context of this thesis, transferable or interorganisational practices are used synonymously, those *transferable or interorganisational practices* are defined as practices that can be carried out by the Hidden Champions and their supply chain partners, either collectively or individually, to mediate the management and mitigation of supply chain disruptions. Carter et al. (2017) identify knowledge sharing and cooperative communication as interorganisational practices that enable the further development of supply chain capabilities such as Supply Chain Risk Management. Those interorganisational practices are comparable to cooperation in Vendor Managed Inventory (VMI) projects presented by Ivanov et al. (2021) or the purchasing alliances presented by Mohr (2009) in the context of Supply Chain Sourcing. Training and development provide a common knowledge base leading to competent and skilled employees with critical thinking skills that are recognised as of high strategic importance for the Hidden Champions and their supply chain partners. Interorganisational critical thinking skills and knowledge development in return positively impact financial performance, which can avoid ex-ante costs for managing risk and reduce ex-post disruption recovery costs, as investigated by Sheffi and Rice (2005), for the cooperating supply chain members. Supply Chain Risk Management training helps Hidden Champions and their supply chain partners overcome the uncertainty challenges that typically arise from supply chain disruptions identify Kirchoff et al. (2016). After all, they will inspire trust and confidence in the supply chain members' Supply Chain Risk Management activities, state Wong and Karia (2010), which leads to a quick reaction in case of supply chain disruptions. For further use in this research project, training and development of employees are defined as *interorganisational supply chain risk management practices* where the members of an upstream supply chain can share their knowledge on Supply Chain Risk Management practices to mediate the management and mitigation of disruptive supply chain events. Hence, the Hidden Champions' training and development of employees are essential practices of the supply chain practice view. Disseminating knowledge on Supply Chain Risk Management practices in the upstream supply chain administered by the Hidden Champion is a cooperative approach that requires swift and seamless two-way communication and joint learning through Supply Chain Risk Management training, thus contributing to increased resilience by benefiting from the supply chain partners' experiences (Carter et al. 2017). Applying Supply Chain Risk Management cooperation as interorganisational practice is of growing importance for the Hidden Champions, with international supply chains increasing complexity. The advantages of cooperation are demonstrated by Din et al. (2013), identifying excellent innovation governance as a confirmed success strategy. Hence, through intense cooperation with their customers, Hidden Champions achieve superior performance (Simon 2009). When understanding the benefits of a cooperative approach to manage and mitigate disruptive events in the Hidden Champions' upstream supply chain, companies participating in such cooperative Supply Chain Risk Management can react more quickly to disruptive events. As a result, according to Walker et al. (2000), once the potential of cooperation is unlocked, the supply chain's reliability increases, and the performance of its processes improves.

3.4 Chapter Summary

This chapter, Chapter 3, presented the theoretical framework to this research based on the supply chain practice view of Carter et al. (2017), spanning the organisational boundaries of the Hidden Champions and embracing the partners in the upstream supply chain. This theoretical approach allows the author to identify how interorganisational cooperation supported by training and development is disseminated between the Hidden Champions and their upstream supply chain partners, thus improving the joint Supply Chain Risk Management process. The next chapter, Chapter 4, presents the research methodology for this thesis which builds on this theoretical framework and aims to understand the Supply Chain Risk Management Processes used at the Hidden Champions and how they cooperate with their supply chain partners.

Research Methodology 4

Within the previous chapter introducing the theoretical framework of this study, the theoretical foundations that helped to frame and guide this research project were discussed. These theoretical foundations create the basis for the research methodology applied to this research project. This research project aims to understand and explain the Supply Chain Risk Management processes used by the Hidden Champions and the area of responsibility in which they are implemented. To this end, it is essential to understand what training and development the Hidden Champions' employees receive to correctly apply the Supply Chain Risk Management processes. Furthermore, it is essential to understand how the communication with the supply chain partners is conducted and what training and development are provided to the supply chain partners by the Hidden Champions to support their Supply Chain Risk Management processes. The author addresses these aims by adopting the supply chain practice view, introduced by Carter et al. (2017), which is the theoretical lens that supports this research project by taking an interorganisational perspective.

This chapter, Chapter 4, introduces the conceptual framework relevant to this study. This study's philosophical position, including the ontological, epistemological, and axiological positions, is discussed in depth, followed by the research design to introduce the enquiry of qualitative data to address the previously defined research question. Given research with human participants, the ethical requirements in the context of this research project are discussed before introducing the appropriate research methods. The section on research methods reflects on the experiences of this project's pilot phase that aimed at testing the findings

Supplementary Information The online version contains supplementary material available at https://doi.org/10.1007/978-3-658-50453-3_4.

© The Author(s), under exclusive license to Springer Fachmedien Wiesbaden GmbH, part of Springer Nature 2026

M. Mayer, *Spanning Organisational and Geographical Boundaries*, https://doi.org/10.1007/978-3-658-50453-3_4

of the prior literature review and builds on them to outline the research methods of the main phase of this study. This main phase consists of qualitative data collection using semi-structured interviews to understand the interorganisational cooperation of Hidden Champions and their supply chain partners to manage and mitigate disruptive events in the supply chain. A focus group discussion (FGD) validated the results of the later data analysis. To test this research process, the author regularly debriefed with his research supervisors and used peer review that provided additional criticality, as proposed by Hail et al. (2011). Subsequently, the coding and analysis of the qualitative data collected in this study's fieldwork are introduced. The results derived from this data analysis will be reported, and the relevant findings will be discussed in the following chapter. To further define the layout of this chapter, its structure is presented below (Fig. 4.1).

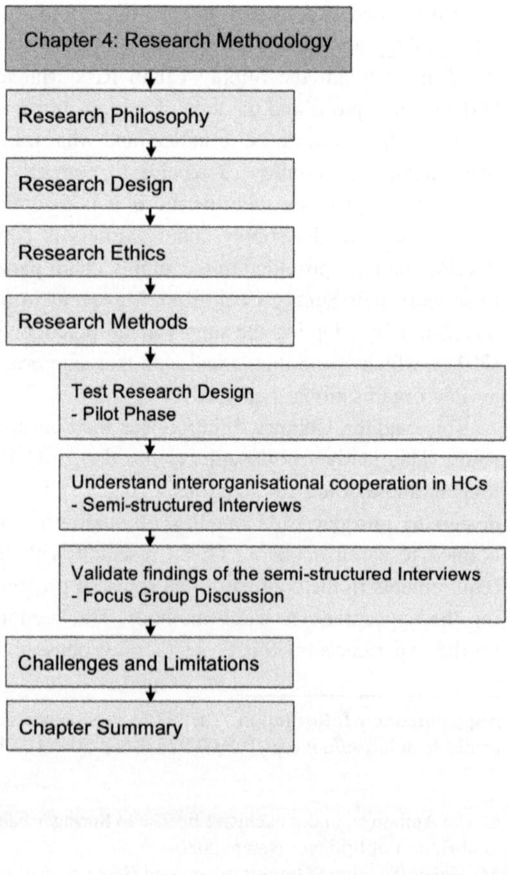

Fig. 4.1 Structure of Chapter 4: Research Methodology

4.1 Research Philosophy

The overarching term of research philosophy describes the development and nature of knowledge (Saunders et al. 2009). Understanding the relevant research philosophies and their philosophical and theoretical underpinnings are essential for every researcher before laying out the research design for their work. Lincoln and Guba (1985) recommend that the researcher, before entering a discussion on research philosophies, identifies his paradigm or the "philosophical world-view" (Creswell 2014, p. 5), also supported by Saunders et al. (2009), defining a paradigm being

> *"a way of examining social phenomena from which particular understanding of these phenomena can be gained, and explanations attempted"* (Saunders et al. 2009, p. 118).

The researcher's access to the knowledge of the social phenomenon shall be researched, together with how this phenomenon can be identified to demonstrate the knowledge created. The researcher's comprehension and understanding of the nature of reality existing within his study and how this knowledge can be apprehended establishes the research design. The following section will discuss these philosophical foundations, commencing with a definition of the ontological, epistemological, and axiological assumptions, and then introduce the research design for this study. Before laying out the research design, understanding the philosophical foundations relevant to this research is required; at the same time, each philosophical foundation will influence the author's thoughts on the research process in a unique way (Saunders et al. 2009). A taxonomy of research paradigms provided by Burrell and Morgan (1982) identifies the four paradigms functionalist; interpretive; radical humanist; and radical structuralist for the analysis in business and management research. The research philosophies drawn from this taxonomy for business and management research are compared, this comparison helped the author to identify the appropriate philosophical position according to the research question. For a better understanding of these philosophical foundations displayed, the author defines the ontological, epistemological, and axiological assumptions in the following section.

The overview in Table 4.1 depicts and compares the philosophical foundations and their individual characteristics in the light of individual research paradigms for business and management research.

Table 4.1 Research Philosophies. (Adapted from Saunders et al. 2009, p. 119)

Research Terms and Definitions	Positivism	Realism	Interpretivism	Pragmatism
Ontology: The researcher's view/fundamental beliefs of the nature of reality	External, objective, and independent of social actors	Objective. Exist independently of human thoughts and beliefs or knowledge of their existence but is interpreted through social conditioning	Socially constructed, subjective, may change, multiple realities	External, multiple views chosen to enable answering of research question
Epistemology: The researcher's view what constitutes acceptable knowledge	Only observable phenomena can provide credible data, facts. Focus on causality and law-like generalisation, reducing phenomena to simplest elements	Observable phenomena provide credible data, facts. Insufficient data means inaccuracies in sensations (direct realism). Alternatively, phenomena create sensations that are open to false interpretation (critical realism). Focus on explaining within a context or contexts	Subjective meanings and social phenomena. Focus on the details of a situation, a reality behind these details, subjective meanings, motivation actions	Either both, the observable phenomena and subjective meanings can provide acceptable knowledge dependent upon the research question. Focus on practical applied research, integrating different perspectives to help interpret the data
Axiology: The researcher's view on roles and definitions of values in research	Research is undertaken in a value-free way. The researcher is independent of the data and maintains an objective stance	Research is value-laden; the researcher is biased by world views, cultural experience, and upbringing; these will impact on the research	Research is value-bound; the researcher is part of what is being researched, cannot be separated from it, and there-fore will be subjective	Values play a key role in interpreting results, with the researcher adopting both objective and subjective points of view
Data collection: Data collection techniques most often used	Highly structured, large samples, measurement quantitative, but may also be qualitative	Methods chosen must fit the subject matter, quantitative or qualitative	Small samples, in-depth investigations, qualitative	Mixed or multiple methods designs, quantitative and qualitative

4.1.1 Ontological Assumption

The researcher's ontological assumptions or ontology is his fundamental belief in the nature of reality. The two aspects of the ontological assumptions are objectivism in contrast to subjectivism. In objectivism, social entities exist independently of social actors. Their social reality is straightforward; social interpretation is not required. In subjectivism, reality is socially constructed (Bryman and Bell 2015), created by the actors of the relevant social entity. To understand this social reality or the meaning that the social actors attach to a phenomenon, interpretation is required to explain organisational life as it is (Saunders et al. 2009). Research philosophies with an objective ontology, such as positivism or realism, and also pragmatism as a research philosophy leaning toward an objective ontology, are primarily external and independent (Saunders et al. 2009) from the research subjects. Therefore, those research philosophies aim to collect data and facts independently from the research subject. For this research project, the author adopts a subjective ontological assumption. This research project takes place at Hidden Champions, primarily family-owned businesses in different business sectors that each attribute a specific business strategy to their organisation; they are the social entities of this research project. Hence, the social phenomenon of Supply Chain Risk Management mediated through interorganisational cooperation is a reality that is socially constructed by the actors of the different social entities.

4.1.2 Epistemological Assumption

The epistemological assumptions or what constitutes acceptable knowledge are essential to understand this research project's social reality. It is necessary to differentiate between data and feeling to answer the question of what constitutes acceptable knowledge. A synonym used for the research paradigm is the "philosophical worldview" (Creswell 2014, p. 5) guiding the researcher. Among the many research paradigms, the two most differentiated paradigms are the positivist worldview and the interpretivist worldview. Within the positivist paradigm, only observable phenomena that can be quantified can provide credible data. The research focuses on cause-and-effect generation and law-like generalisation found in natural sciences (Saunders et al. 2009). Antagonist to the positivist orthodox (Bryman and Bell 2015) is interpretivism. Interpretivism is of high relevance for business and management research, gaining an understanding of and explaining the organisational life as it is (Saunders et al. 2009). This explanatory approach allows for research, especially in fields previously not investigated, such as organisational behaviour, within a complex business situation where data is acquired by interpreting information received from the research participants directly involved in the social event (Saunders and Lewis 2012). See Fig. 4.2 that represents the visualisation of the continuum of research paradigms.

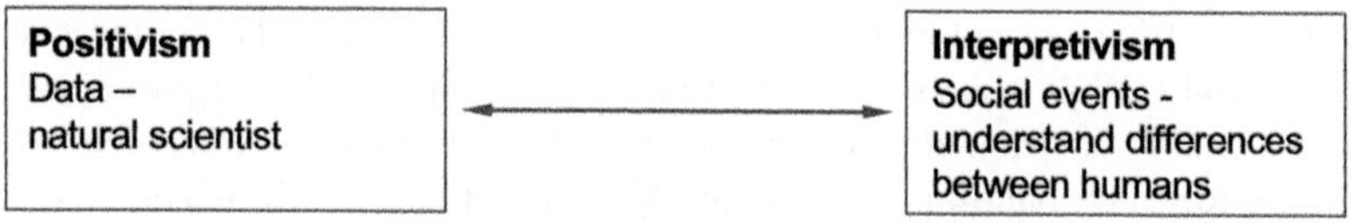

Fig. 4.2 Continuum of the Research Paradigms

4.1.3 Axiological Assumption

The axiological assumptions are the third of the philosophical foundations discussed, which describes the author's views on roles and definitions of value in research. Within the two extrema of the research paradigms' continuum, the positions discussed earlier in this thesis, the positivist researcher is independent of the data and holds an objectivist position. In contrast, interpretivist researchers are part of what is being researched; they cannot separate themselves from the research subject and hence will hold a subjective position.

4.1.4 Interpretive Approach to Research

The above discussion indicates that the interpretive approach is most appropriate when undertaking research based on people and information needs, which is reinforced by this research's question, being a how question that requires an explanatory research paradigm according to the classification of research paradigms defined by Burrell and Morgan (1982). Hence, according to Saunders et al. (2009), an interpretive approach is required to understand and explain organisational life from an experience-near perspective. This research's subjects are senior managers of Hidden Champions in the mechanical engineering industry in the DACH region; each of their companies have multi-faceted business strategies. It is therefore vital to choose a methodological approach that is explanatory in nature. The conceptual framework used in this research allows for interpreting information on social phenomena, for example, Supply Chain Risk Management in complex organisations such as the Hidden Champions (Saunders and Lewis 2012); by entering their natural setting, qualitative data will be collected. To bridge the methodological gap in Supply Chain Risk Management research raised earlier by Sodhi et al. (2012) and generate theory on Supply Chain Risk Management spanning organisational boundaries of Hidden Champions, an explanatory research approach is required, found in the interpretive research approach. This chosen interpretive research approach also extends the scientific knowledge by using a philosophical approach that is not common in Supply Chain Risk Management research (Darby et al. 2019), with most research in this discipline adopting a quantitative approach (Okongwu et al. 2015). The reliability and validity of the interpretive approach are explained in the following section, which introduces the research design of this study.

4.2 Research Design

The philosophical framework identified in the previous section of this thesis allows for interpreting information on social phenomena (Saunders and Lewis 2012), for example, Supply Chain Risk Management in complex organisations such as the Hidden Champions in the DACH region. For research in business and management, especially in fields such as organisational behaviour within complex organisations, according to Saunders and Lewis (2012), interpretive research is of high relevance. This interpretive research is in contrast to the tradition of the natural scientist (Bryman and Bell 2015), in which the majority of research in this discipline is positioned, as stated by Okongwu et al. (2015). Since there is a methodological gap in Supply Chain Risk Management research stressed by Sodhi et al. (2012), and to the author's knowledge, there is no research on Supply Chain Risk Management spanning the organisational boundaries of Hidden Champions in the DACH region, a qualitative, theory-generating research approach is required. By entering the natural setting of the Hidden Champions, the author will collect qualitative data on the existing Supply Chain Risk Management approaches, which will help to develop new theory based on interpreting qualitative data (Gioia et al. 2013). Based on the supply chain practice view introduced by Carter et al. (2017), applying Supply Chain Risk Management needs to span organisational boundaries to embrace the partners in the Hidden Champions' upstream supply chain. This requires imitable or transferable practices that all supply chain members can use. These transferable practices are translated from the Hidden Champions' success strategies of customer proximity and innovation activities as introduced by Simon (2009) in the initial literature review of this thesis. Based on the identified Hidden Champions' success strategies, the close cooperation with supply chain partners and interorganisational networks lead to a better understanding of the relevance of Supply Chain Risk Management for the success of the Hidden Champions and their upstream supply chain partners. Ultimately, based on these findings, the research question of this study is:

> **How does interorganisational cooperation between Hidden Champions and their supply chain partners mediate the management and mitigation of disruptions occurring in the upstream supply chain of Hidden Champions in the DACH region?**

This research project follows the interpretivist philosophical tradition, which is interested in understanding the social reality or the meaning that the social actors attach to a phenomenon. Therefore, interpretation is required to explain organisational life as it is (Saunders et al. 2009). This research will incorporate methods drawn from qualitative approaches to data collection and analysis. A two-step pilot phase was introduced to test the research design and support this qualitative research's validity. After reflecting on the learnings of the prior pilot phase, semi-structured interviews with senior managers responsible for Supply Chain Risk Management, located mainly in purchasing or supply chain management, are introduced. Coding and data analysis following the systematic approach of Gioia et al. (2013) provide rigour to this qualitative research. This study employs a focus group discussion to inform the interview partners of the findings from the previous interviews, hence validating whether the data analysis' findings of the semi-structured interviews correspond to the interview partners' expressions. Since focus group discussions are particularly likely to prompt group interaction and subsequently concentrate on the group's attitudes, understanding, language, and group norms, such as informal behavioural guidelines (Kitzinger 1994), it helped the author to provide a more in-depth discussion. While the presented research design enters into research with human participants, the ethical requirements of Bournemouth University are shown in the next section.

4.3 Research Ethics

This thesis is fully compliant with the requirements of the Bournemouth University Research Ethics Committee (REC) when entering into research with human participants. Research ethics aim to design, review and implement the research process to ensure the integrity, values and quality of research to add value to society and contribute to knowledge. According to Flick (2018), a primary ethical concern in qualitative research is human participation. Researchers need to identify vulnerable individuals and protect them from harm. In this study, the research participants were senior managers of Hidden Champions, responsible for Supply Chain Risk Management, located mainly in purchasing or supply chain management, so there is an independent and equal relationship between the research participants and the researcher. Participation in this study was entirely voluntary, and it was possible to withdraw participation at any time; upon withdrawal, all data collected was to be deleted. The data collected during this research project was anonymised for further use in this study and the later evaluation of this thesis. The data is stored on a password-secured server for five years after finalising this thesis. To receive ethical approval, all participants in this study were to receive information about the study itself and had the opportunity to give written consent or reject their participation in this study at any time. This thesis's ethical concern was the anonymity of the study participants, their businesses, and the related products. These products are unique and could be used as a trace back to the business they were produced at. All participants of this study have given their informed consent but requested to remain anonymous throughout the study and in later publications. Upon request of the interview partners, the author anonymised all names, products, and businesses by using pseudonyms, similar to the advice of Saunders et al. (2009) as presented in Table 4.5 Interview pseudonyms. Following the requirements of Bournemouth University's Research Ethics Committee (REC 2021) the forms obligatory for the ethical approval, including Participant Information Sheet (The relevant data can be found in Online Resource 1), Participant Agreement Form (Online Resource 2), and the corresponding interview guidelines (Online Resources 3; 4 and 6) were sent to Bournemouth University's Research Ethics Panel Chair for approval prior to performing the various data collection stages. The Covid-19 pandemic invoked several modifications to the research design; ethical approval for this study, including the relevant amendments for a personal and remote focus group discussion was obtained. The approved documents from the Ethics Panel Chair are available on request.

4.4 Research Methods

Research methods describe the procedures to collect and analyse data for this research that are suitable to answer the research question. This section starts with the reflection and learning from the pilot phase of this research project that served to assess the research design's feasibility, tested the interview set-up and adapt the research design, if necessary.

4.4.1 Reflection and Learning from the Pilot Phase

Using the pilot phase, the author chose a two-step approach to support the qualitative research process' validity, test the interview set-up, and reduce bias recommended by Saunders et al. (2009) and Christopher et al. (2011). This two-step approach enabled the author to adjust the interview guideline and allowed for feasibility testing of the research design itself, question validation and ensuring the reliability of the information collected following the suggestions of Saunders et al. (2009). Pilot tests are often used in research to ensure good practices and represent a fundamental step in planning research. For the author, who was inexperienced in qualitative research, the recommendation of Leon et al. (2011) to undergo a pilot phase was beneficial. It helped the author to learn about the challenges of qualitative research, build the necessary competencies and skills from these learnings, and confirm the literature review findings (van Teijlingen and Hundley, 2002).

The first step of the pilot phase commenced with an expert interview to validate the findings of the initial literature review manifested in the interview guideline used for the expert interview. The expert identified for the interview is a profound practitioner with educational experience in purchasing and supply chain management. Based on long-standing business relationships with Hidden Champions from the DACH region, the expert is seen to hold an objective and an open position in the topic area (Helfferich 2014) of Supply Chain Risk Management in Hidden Champions that should support the aims and objectives of this research project. The expert interview took place in a telephone interview and provided rich topic-centred information when discussing and validating the literature review's findings. The interview guideline prepared for this expert interview, which builds on these findings from the literature review, is included in Online Resource 3. The interview guideline gives the reader insight into the development of the questions prepared for the first interview guideline, elaborated in great detail and including probing questions that helped the author validate the expert's answers. Based on the experience from the expert interview, the author prepared for the pilot interview discussed in the next section.

The second step comprised a pilot interview with a senior manager in purchasing and supply chain management of a German Hidden Champion manufacturing products for the consumer goods industry. The pilot interview took place on the premises of the Hidden Champion; the interview partner suggested the staff restaurant as the location for the interview. This setting was unique in many ways and allowed for an open and relaxed atmosphere where the interview partner felt most comfortable, despite being under time constraints. Thus, the author needed to plan the general timing for the field research thoroughly while the willingness of interview candidates to participate is enormously dependent on the timing of the interviews during the year; hence these interview dates need to be planned well ahead. Moreover, the interview provided extensive verbal and non-verbal information; therefore, planning needs to consider a limitation of the interview duration to avoid the excessive time required for transcription later on.

4.4.2 Purposeful Sampling Strategy

The focus on the Hidden Champions in the mechanical engineering industry in the cross-border area of the DACH region is evident, given their high economic relevance and geographic concentration discussed earlier in sect. 2.3.1. Given their global market-leading position, a high degree of internationalisation of the firm is sought to provide rich insights into the supply chain complexity of the Hidden Champions. Therefore, the author used a purposeful sampling strategy inspired by Poulis et al. (2013) and Stuart et al. (2002). Purposeful identification of samples, instead of random sampling to achieve empirical generalisability, is recommended by Saunders et al. (2009) for qualitative research; it served the research the most by achieving an information-rich contribution by the selected companies. Since it was not evident to the author what the optimal number of interview partners was when following Gioia et al. (2013), he adopted Eisenhardt's (1989) recommendation to limit the number of interview partners to between four to ten. Once the author repeatedly identified similar themes in succeeding interviews, "theoretical sufficiency" was reached, the research stopped (Diaz Andrade 2009, p. 51). The ten interview partners' selection followed a cross-sectoral approach to reduce the risk of an industry sector-specific bias. A table with potential interview contacts was developed, an anonymised excerpt of which can be found in Table 4.2.

Table 4.2 Excerpt of Potential Interview Contacts

Company	Contact / Function	Email	Street	ZIP	Place	Country	Contact made
Wxxxxx	Mxxx Jxxxxxxxxx / Human Resources	Mxxx. Jxxxxxxxxx@w xxxxx.com		xxx xx	Mxxxx xxx	DE	
Mxxxxx	Mxxxxx Vxxx / Supply Chain	Mxxxxx.Vxxx @Mxxxxx.de		xxx xx	Hxxxx xxxxxx	DE	
Zxxxxx	Pxxxx Dxxxxx / Purchasing	Pxxxx.Dxxxxx @Zxxxxx.xx		xxx xx	Lxxxx xxxxx	DE	
Exxxxxx Axxxxxxx	Mxxxxxxx Xxxxxxxx /CEO	Mxxxxxxx. Xxxxxxxx@ Exxxxxx.xx	Xxxxxxst rasse xx	xxxx	Rxxxx xx	CH	Mail 201909 20
Axxxxx Gxxxx	Hxxx Xxxxxxxxx / CEO	Hxxx.Xxxxxxx xx@ Axxxxx-Gxxxx.xxx	Xxxxxxxx xstrasse xx	xxxx	Lxxxx xxxxx	CH	Mail 201909 20
...	...	...	...	...	...	...	...

The Hidden Champions in the DACH region relevant for this study were identified based on the *World Market Leaders Index* (Müller 2021), an online database offering objective and transparent selection criteria for global market leaders and were chosen based on their affiliation to the mechanical engineering industry in the cross-border area of the DACH region. The publisher of this database is the University Entrepreneurship Corporation gGmbH (Müller 2021) which seeks to foster research on global market-leading businesses in the DACH region. Based on the most recent definition for Hidden Champions given by Simon (2021), they are global or regional leaders in their market segment; with revenue below €5 billion and a low level of public awareness that often can be found in family ownership. The potential interview partners relevant for this study were identified based on company information available online. The author personally contacted those potential interview partners via telephone, email or post to explain the benefits and objectives of this research and the estimated amount of time planned for the research process. Where it was not possible to identify direct contacts in purchasing or supply chain management, the author sought direct contact with the Hidden Champions' owners or managing partners. In a personal letter, the Hidden Champions' owners or managing partners were informed about the background and benefit of this research project for their company and asked for help identifying a contact person. Official permission to conduct research in the company was obtained through these contacts by following the considerations on gatekeepers of Ahern (2014). Given the Hidden Champions' adversity to publicity, this study builds on the experience of Okumus et al. (2007) to build a trustful relationship to gain access for research in complex business environments. Access or rejection were experienced already during the selection of interview partners in the pilot phase of this research, reflected in neglection or rejection of the invitation, or questioning the relevance for research on Supply Chain Risk Management in Hidden Champions. Whenever the author did not receive any feedback on his invitation to participate in this research project, he followed up if there was any interest in participating in this study. Where there was no interest, he asked if it was possible to enquire about an appointment later. He put the appointment on hold and called again at the specified date if the answer was affirmative. If there was no interest or the appointment was cancelled immediately, the author did not pursue this contact further. Due to this complexity and the identified lack of Supply Chain Risk Management in Hidden Champions, it was essential for the author not to appear insensitive. The contact and the later challenging enquiries about the existence of Supply Chain Risk Management in the participating businesses needed to be made in a very diplomatic way not to lose the trust placed in the author by the interview partners following Maier and Monahan (2010). To avoid rejection by the interview partners and build trustful relationships to gain access to these businesses, the author granted anonymity of their person and function and the participating businesses themselves to all interview partners. Furthermore, this diplomatic approach supported the honest, open and comprehensive qualitative data acquisition during this research which Bradburn et al. (2004) recommend. To meet this requirement, the author has structured the resulting anonymisation in a detailed presentation with the corresponding pseudonyms discussed in sect. 4.4.6.

4.4.3 Data Collection using Semi-structured Interviews

In this study's main phase, the author interviewed the previously selected senior managers of the Hidden Champions responsible for Supply Chain Risk Management, located mainly in purchasing or supply chain management, using semi-structured interviews to collect qualitative data. Therefore, the author defined an interview guideline addressing this project's research and sub-research questions. When interaction with the interview partner and the adaption of the interview questions to examine a particular issue is needed, semi-structured interviews are a highly suitable instrument advise Saunders et al. (2009). Due to this instrument's flexibility in acquiring qualitative information from the field, the author obtained detailed processual and contextual knowledge, as Meuser and Nagel (2002) indicated. Based on the interview guideline, the semi-structured interviews helped the author to gain topic-centred information based on the interview partners' experience and enabled information-rich discussions with the interview partners (Helfferich 2014), expanding the author's knowledge for the following interviews. The interview guideline used consisted of broad questions focusing on practices to manage and mitigate risk in the supply chain in the organisation of the Hidden Champions. The interview guideline helped to avoid bias during the research phase, with questions that considered facilitating critical verification on the one hand and avoided leading the interview partner on the other; once the author posed such questions in the heat of the discussion, they were marked during the interview and put on a "Not-To-Do-list" for the following interview. Being open to adapting the interview guideline also allowed to act flexibly during the following interviews, where the interview partners' guidance opened new perspectives that enriched the following interviews. Therefore, the author sought personal meetings with the interview partners; wherever this was not possible due to logistical or pandemic reasons, this study used electronic communication, however being aware of the limitations of electronic communication that will be reflected on in detail in sect. 4.5 "Challenges and Limitations of this Research Design".

A tabular overview of the timeline of this research project, including interview dates, the interview partners' countries, the assigned pseudonyms, and the interview partners' positions in their organisations, can be found in Table 4.3.

Once the author had established contact, interview dates and locations were agreed upon based on the availability of the interview partners. Most interview partners preferred a face-to-face meeting at their company headquarters; two interviews were conducted via telephone due to the interview partners' business constraints. All the interviews were audio-recorded, and notes were taken during the discussion. The evaluation of these interviews took place directly afterwards based on the notes made during the interview, which informed the following interviews. In this way, the author could incorporate newly acquired knowledge into the following interviews.

Table 4.3 Timeline of this Research

Interview Date	Country	Pseudonymous	Interviewee position	Remarks
17 Jul. 2018	Germany	**Expert**	Director Materials Management	Telephone
21 Sep. 2018	Germany	**Pilot**	Director Purchasing	Face to Face
11 Jul. 2019	Germany	**Evacuate**	Head of Purchasing	Face to face
16 Jul. 2019	Germany	**Experiment**	Director Global SCM Head of Purchasing	Face to face
13 Aug. 2019	Switzerland	**Conche**	Chief Procurement Officer	Face to face
11 Sep. 2019	Austria	**Lift**	Chief Operating Officer	Face to face
20 Sep. 2019	Germany	**Mill**	Head of Logistics and Purchasing	Face to face
30 Sep. 2019	Austria	**Open**	Corporate Insurance & Risk Management	Face to face
08 Oct. 2019	Austria	**Glide**	Head ofPurchasing & Logistics	Telephone
15 Oct. 2019	Germany	**Irrigate**	Director Global Procurement	Face to face
21 Oct. 2019	Switzerland	**Travel**	Head of Procurement	Face to face
30 Oct. 2019	Switzerland	**Cooper**	Vice President Logistics & Manufacturing	Telephone
05 May 2020	Online	**FGD**	See above	Online

4.4.4 Coding and Data Analysis

Coding and data analysis of the qualitative data transcribed and translated for this research were based on the systematic approach of Gioia et al. (2013) that provided qualitative rigour, delivering plausible and defensible conclusions. This systematic approach supports process research by allowing the author to deeply immerse in the topic using the original terms and expressions of the interview partners, which supported the ample depiction of their real-life experiences. Joining the interview partners' terms and expressions and the author's theoretical topic areas led to developing new models. Initially developed for the analysis of qualitative interviews, the Gioia Method is highly suitable for coding and analysing group interviews or discussions such as this research's focus group discussion. Data analysis took place in a multi-stage process in parallel to the interviews following the recommendations of Gioia et al. (2013). The coding was done after the interviews by relistening the interview recording while in the car on the way back to the author's office and doing an initial coding in writing, or when sitting on a train, relistening and coding simultaneously. Furthermore, the data collected during the first interviews informed the later research process and guided the interview questions that were asked in the later interviews. This interrelation of data collection and data analysis is seen as one of the strengths of this qualitative approach to research, as confirmed by Ezzy (2002). The method chosen only recently gained acceptance in Supply Chain Risk Management research and supports the understanding of processes and the corresponding contexts in the interview partners' companies. It also supported the later focus group discussion that aimed to validate the findings of the initial data analysis. The focus group served further to collect qualitative data and gain a rich narrative of the focus group discussion partners and their experiences of organisational life (Gioia et al. 2013). The benefits of the adopted method are its imminent flexibility to adapt the research as it progresses. The interview guideline was adapted based on newly occurring topics during the research process; according to Saunders et al. (2009), this adaption of the interview guidelines are prerequisites embedded within interpretive research that allowed to uncover and develop new topics during this research project. This method is also advantageous compared to other qualitative research methods, such as the probably most popular qualitative research method, case study research articulated by Yin (1984) or Eisenhardt (1989). Advantageous in a way that the Gioia Method (Gioia et al. 2013) provides a real-life picture of the interview partners' experience and supports the transition from data to theory through its graphical representation of new theories (Gehman et al. 2018), demonstrating rigour in qualitative research. The revelatory potential of this qualitative research through the intimacy between the interview partners and the author (Gioia et al. 1994) led to a rich narrative that needed to be transferred into a textual form to use the qualitative data collected during the field research for coding and analysis. The following section provides information on the transcription and translation of the collected data for further coding and data analysis used in this research project.

4.4.5 Transcription and Translation

The semi-structured interviews and the focus group discussion being the heart of this research were audio-recorded; using this qualitative data collected for the necessary coding and analysis was only possible by transferring the interview partners' original language into textual form. When collecting data, the author took notes on newly emerging topics or non-verbal communication that one could not identify by listening to the audio recordings. Saunders et al. (2009) recommend relistening the recordings immediately after each interview to deeply immerse in the interview partner's narrative. While doing so, the author commenced transcribing and annotating the audio recordings to ensure the transcriptions' accuracy supporting the qualitative analysis and using the interview partners' verbatim statements as direct quotations in this thesis (Saunders and Lewis 2012). Following transcription, the author translated the interview partners' original language into English to allow for usage in this research project. Given the solid dialectal colouration of some of the interview partners, it was necessary to listen to the audio recordings several times to guarantee the reproduction of the verbatim statements in standard English within the transcripts. However, the German-speaking interview partners used strong local accents from southern Germany, Austria and Switzerland. To avoid losing the meaning of the interview partners' original narrative during translation, the author focused on most meaningful translating of the "original voice" of the interview partners. Transcribing the interview with Travel was very time-consuming due to the local terms used. For example, his approach to mitigating the effects of a supply chain disruption at one of TRAVEL's strategic supply chain partners by motivating their employees; therefore, Travel explained he was rewarding a "Znüni" (Travel 56) to the employees of the disrupted supply chain partner. Znüni is Swiss-German dialect for a snack eaten at nine o'clock; the respective German term "zu neun Uhr" is then moved together into Znüni. An English equivalent would be elevenses, a tea or coffee break accompanied by a light snack at eleven o'clock (Jaine 2014). Already during the earlier pilot phase, the author acknowledged the duration of each semi-structured interview and its later impact on the duration of the transcription process. Due to the author's lack of experience in transcribing interview data, he tested different transcription forms in the pilot phase to identify the best possible transcription form that could be used for later data analysis. The author's own manual transcription was very time-consuming, requiring a minimum one hour of transcription time per ten minutes of interview time, depending on the audio recording quality and the clear pronunciation and rhetoric of the interview partner. This manual transcription stood in contrast to an automated transcription engine, which transcribed the audio recordings of the interviews and made them available in the form of a text document. Despite the cost of this automated transcription and necessary proofreading required, this form turned out to be the most efficient, allowing extensive audio recordings to be transcribed on time and in decent quality. It also freed up time, which would otherwise have been needed for manual transcription, and saved the cost for a professional transcriber who needed to swear to confidentiality. After transcription, the author relistened all audio recordings to finalise the transcripts and compensate for the automatic transcription's disadvantages inhibiting complete immersion in the interview data. Following transcription, the author compared the transcripts with the notes taken during the interview to enhance the transcripts' quality further. Time stamps attached to the quotes supported orientation in the transcript that was required when relisting the audio recordings during the comparison and annotating the transcripts. During this relistening, once noteworthy situations from the interview were recalled, those were added to the transcripts to support the documentation of the research results. For better reader guidance, the participants' verbatim statements were treated as direct quotations in this thesis. Therefore, the interview partners' statements, followed by the individual identifier, are shown in italics; necessary insertions were made in squared brackets and are shown in non-italics. Thus, the first code of the first semi-structured interview with Evacuate received the identifier "Evacuate 1". Furthermore, if the question to be answered did not emerge from the context discussed, it was included prior to the answer in the direct quotation. Hence, the full code used in the presentation and discussion of the findings has the following appearance, see Fig. 4.3:

> [How is the SCRM process referred to in your company?]
> "*Good question* [pause]?" (Evacuate 1)

Fig. 4.3 Example of Interview Question and Coding

This enabled the statement's actual content also to be presented to the external observer in an easily comprehensible way. Also, the interview partners' verbatim statements were smoothened, and the interview partners' repetitions were removed to increase further readability and usage of the transcripts in the analysis. Where statements or sentences were not complete or where it was necessary to smoothen out sentences, certain words were added in square brackets or inserted omissions marked with .... Similar was done for repetitive words, which were removed as shown in the following example, see Fig. 4.4:

> "*We experience this liveliness-factor in several audits ... where you can't bullshit* [sic] *them* [the auditors] *anymore.*" (Expert 29)

Fig. 4.4 Example of Smoothened Coding

This smoothing of sentences enabled the interview partners' verbatim statements to be easily understood without altering their content or meaningfulness (Table 4.4).

Table 4.4 Excerpt Coding Table

Informant Code	Informant's Original Expression	Translation	1st-order Concepts Informants Viewpoint	2nd-order Themes Theory-based Concepts	Aggregated Dimensions
T88	[Training für die Mitarbeiter?] Nein, haben wir eigentlich nicht. Training und so haben wir nicht.	[Training for employees?] No, we don't really have it. Training and so we don't have.	SCRM Training does not happen	SCRM Training	SCRM Culture
T94	[Training im Bereich Beschaffung] Gibt's ja, was es gibt. Wo wir Training jetzt haben ist zwar nicht mein Lieblingsthema aber Compliance wurde geschult.	[Training in procurement?] Is there, yes what is there: Where we have training now, is not my favourite subject, but compliance has been trained.	Compliance training is existing	SCRM Training	SCRM Culture
G85	Eher on the Job. also eine eigene Schulung dafür haben wir nicht, nachher dann im Zuge der Einarbeitung oder vom Kollegen nachher.	Rather on the job. We don't have our own training for this, but later in the course of introductory training or from colleagues.	Own training approaches on the topic of complaint handling / SCRM are not available	SCRM Training	SCRM Culture
M84	[SCRM Training für die MA?] Einkäufertagung einmal im Jahr, da kommen alle Einkäufer weltweit zusammen, da ist das Thema Risk Management immer ein Thema.	[SCRM Training for the Employees?] Buyers conference once a year, where all Buyers worldwide come together, Risk Management is always a topic.	Internal collaboration within purchasing, but no formalised trainings offers.	SCRM Training	SCRM Culture
C85	Es gibt keine internen Schulungseinheiten. Es gibt Schulungsansätze	There are no internal training units. There are training approaches	SCRM Training is available on rudimentary level	Learning and training to be improved	SCRM Culture

The participants of the semi-structured interviews each were given their individual identifiers based on their correlating pseudonyms. In this way, the author could easily identify the individual informant's original expression in the coding table above during the later coding process. For the narratives of the focus group discussion used in the presentation of findings, each of the codes also received an explicit identifier using the abbreviation FGD, followed by a consecutive number and in brackets the pseudonym of the participants (Fig. 4.5), which reads as follows:

"And at some point, you told us exactly this, this geo-redundancy among suppliers. To be honest, I didn't discuss that either." (FGD 29 (Lift))

Fig. 4.5 Example Focus Group Discussion Coding

Following the examples of the interview partners given above, the succeeding section presents the anonymisation of the interview partners, their business names, and the products manufactured to allow utilisation for later data analysis and discussion without compromising the guaranteed anonymity.

4.4.6 Anonymisation of Interview Partners

The interview partners in this research project, senior managers responsible for Supply Chain Risk Management, located mainly in purchasing or supply chain management within Hidden Champions, were guaranteed anonymity in this study's published results and in the data that was to be archived. Following the aim of anonymity, the interview transcripts used the interview partners' pseudonyms to describe the product or organisation named in the interviews to guarantee the anonymity requested by the interview partners. In addition to ensuring the anonymity of the interview partners, the businesses' identity was also to remain confidential. Many of these businesses sell unique products, with only scarce competition globally. Consequently, conclusions about the business name and the interview partner could be drawn from the products manufactured. Therefore, similarly to the interview partner, the product is given a pseudonym to avoid recognising and characterising organisations and products. During the finalisation of the transcription process, the author re-read all the transcripts identifying all names, products, businesses, or business sector-related terminology, which then were replaced with the given pseudonyms, see Table 4.5.

Table 4.5 Interview Pseudonyms

Pseudonymous	Product Description
Expert	Autonomous driving systems
Pilot	Leisure-time activity systems
Evacuate	Evacuation and filtering systems
Experiment	Test equipment systems
Conche	Food processing systems
Lift	Heavy goods handling systems
Mill	Non-metallic materials processing systems
Open	Fastening and connection systems
Glide	Outdoor equipment service systems
Irrigate	Commercial irrigation systems
Travel	Powered vehicle systems
Cooper	Water container forming and testing systems

In addition to ensuring the anonymity of the interview partners, the businesses' identity was also to remain confidential. Once the pseudonym was used for identifying the business of the interview partner, the business' name was replaced with the pseudonym written in capital letters, "EXPERT", instead of the Expert's business. Specific names of machines or systems were replaced with "machine" only and written in squared brackets [machine] to identify the changes made to the interview partners' verbatim statements. These changes to the verbatim statements supported the anonymisation of particular machines or systems produced by the interview partners' organisations, which otherwise would be difficult to anonymise. Once the transcripts were prepared, translated, and anonymised as presented above, they could be used to further analyse and structure the data that will be reported on in the next section.

4.4.7 Data Structure

To structure the results of this last step of the analysis and to demonstrate the qualitative rigour inherent to this method (Gioia et al. 2013), the author depicted this study's data structure following (Fig. 4.6).

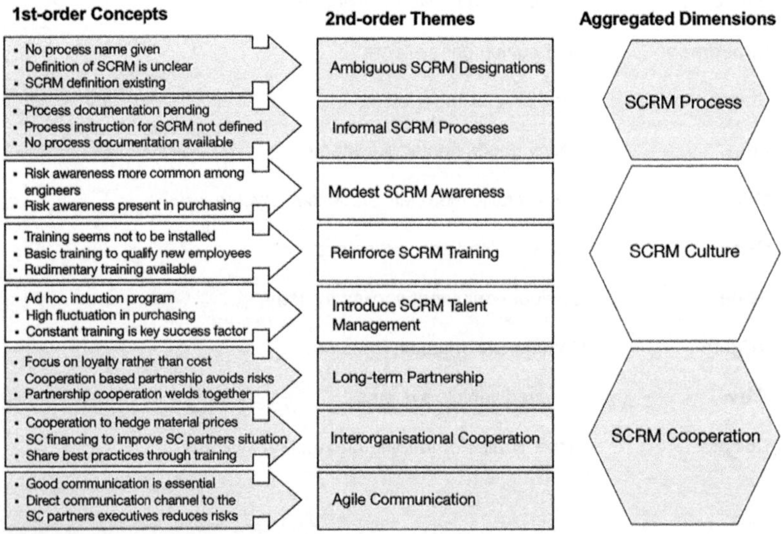

Fig. 4.6 Data Structure of this Research

This first-order analysis was based on the interview partners' original expressions, terms, and descriptions and represented a first explanation of the information received during the interview, which the author titled first-order concepts. These first-order concepts were discussed and tested for similar concepts within one interview and across all interviews, which led to an inflationary development of concepts, however, enabling a new way of thinking derived from the research that would be impeded otherwise by pure replication of the already known. In the second step of the analysis, the author sought similarities between the first-order concepts in light of existing theories that led to second-order themes explaining the phenomena under study. In the last step of the analysis, the theory-based second-order themes were collapsed into top-level, aggregated dimensions in line with existing literature to underpin theorising. This data structure provides an overview of how the author progressed from raw data found in the interview partner's original expressions to the theory-based concepts of second-order themes, which then were combined into aggregated dimensions.

4.4.8 Validate Findings through Focus Group Discussion

The focus group discussion provided rigour to this qualitative study by validating the findings of the data analysis in following Stewart et al. (2007) recommending validation of research findings using focus group discussion. By informing the interview partners about the findings from the data analysis, the author ensured that the data analysis results corresponded with the interview partners' expressions given during the earlier semi-structured interviews. This feedback to the interview partners confirmed that the results derived from that data analysis were plausible and made sense to those experiencing the phenomenon of Supply Chain Risk Management in their business life. While focus group discussions are a well-known qualitative research method, their application ranges from health and social care to market research or politics (Saunders et al. 2009). The terms *focus group discussion* and *focus group interviews* are used interchangeably. While focus group research relies on group discussion, the author will use the term focus group discussion during this thesis. Focus group discussions are particularly likely to prompt group interaction and subsequently concentrate on the group's attitudes, understanding, language, and group norms, such as informal behavioural guidelines, says Kitzinger (1994), which the author sought to provide a more in-depth discussion. Moreover, the author's role as a moderator differed in focus group discussions from that in individual interviews. The author's role in a focus group discussion is to moderate the discussion, putting the author in a reduced position; however, a critical, high-level perspective that strongly supported theorising development and information validation. This high-level perspective also helped the author to avoid disadvantages such as internal bias and falsifying information. Both can arise from groupthink, where participants might agree even if they do not share the opinion or are even intimidated to participate in the discussion. In contrast, some participants tend to dominate the interview, as van Teijlingen and Pitchforth (2006) discussed.

Based on the author's professional experience in organising multi-day events with many participants and the experience from the previous semi-structured interviews on the academic requirements for conducting research with human participation, the focus group discussion was thoroughly planned. At the end of each semi-structured interview, the author asked the interview partners whether they were willing to participate in a later focus group discussion planned for the second quarter of 2020. Most of the interview partners already accepted the invitation after the interview, allowing for the effective composition of the focus group discussion. Precise planning in following the recommendations of van Teijlingen and Pitchforth (2006) took place to find suitable locations accessible to all participants within a one-hour journey. Once the potential hosting organisations confirmed the facilities suitable for a focus group discussion, the author sent an individual invitation including an agenda to each of the earlier interview partners. With the increasing demand for social distancing and travel restrictions prompted by the Covid-19 pandemic in the DACH region, the author transformed the planned face-to-face meeting into an online meeting. Although this offered a straightforward and more cost-efficient solution, it led to more complicated coding and analysis, lacking the perception of body language, which is essential, while attributed to the interview partners' verbatim statements (Collard and van Teijlingen 2016). Following the Bournemouth University's Ethics Panel Chair's approval for this change in research methodology, the author informed the interview partners about the modified framework conditions.

Based on the positive experience when piloting the semi-structured interviews, the author also planned to pilot the focus group discussion. The pilot focus group discussion aimed to test the discussion setting, identify the optimal number of participants, and determine the time required to obtain rich and meaningful information, as van Teijlingen and Hundley (2002) recommended. It should further help the author master the challenge of moderating the focus group discussion, putting the author in a reduced position, and avoiding any participation in the discussion, as advised by McLafferty (2004). This critical, high-level perspective that supported theorising development and information validation also helped the author to avoid disadvantages such as internal bias and falsifying information. Both can arise from groupthink, where participants might agree even if they do not share the other's opinions or are even intimidated to participate in the discussion. In contrast, some participants tend to dominate the interview, as van Teijlingen and Pitchforth (2006) discussed. Given the business implications for managing production stops caused by short-time work combined with supply chain bottlenecks due to the early Covid-19 pandemic, which resulted in the minimal time available from the interview partners, piloting the focus group discussion was no longer perceived as being an acceptable approach. Hence, the author had to carry out the focus group discussion without prior piloting, drawing on his experience from professional practice in moderating larger group discussions and on scholars experienced in using focus group discussion in their research (McLafferty 2004; van Teijlingen and Pitchforth 2006; Collard and van Teijlingen 2016).

With the focus group discussion to take place as an online meeting, both the author's own and the participants' increasing experience working in virtual environments was incredibly supportive, as confirmed by Sarkis et al. (2020). Compared to the in-person focus group discussion, the online meeting also offered advantages by avoiding travel and reducing costs for facilities and catering. More important was the changing group behaviour; initially dominant participants became more involved in each other's conversation flowing from participant to participant through the online meeting format. As put forward by Collard and van Teijlingen (2016), the online meeting provided an uncomplicated way to audio-record the focus group discussion and thus obtained rich data for later analysis. At the beginning of the online meeting, the author referred to previously granted anonymity and confidentiality as requested from the participants in advance in their written consent. Thus, the participants could decide whether they wanted to make themselves known in the discussion or preferred to stay anonymous, hence being addressed with a pseudonym. However, all participants introduced themselves with full names, functions and company names. To support the moderation of the focus group discussion, the author introduced an agenda for the focus group discussion and described the audio recording of the discussion, which was required for later transcription and analysis within this research project. Following the field research results' presentation, the author used a discussion guideline with open-ended questions based on the individual levels of the data structure, as discussed in sect. 4.4.7, as a result of the findings from the data analysis from the semi-structured interviews. This guideline served to ensure the interaction of the focus group discussion participants within the topic area. It helped the author benefit from the participants' rich experiences, which led to generating information-rich data based on their different viewpoints, as Gill et al. (2008) recommended. Since the focus group discussion consisted of German-speaking participants, the author formulated the guideline in German; a translated version to allow for usage in this thesis is included in Online Resource 6.

At the end of the focus group discussion, the participants addressed their strong interest in the findings of this study; hence, the author considered the possibility of further discussion on current topics in supply chain management, especially Supply Chain Risk Management. The focus group discussion can therefore be considered a successful part of this research, enormously helping to deeper immerse in the topic area and validate this research's findings. The next section discusses the challenges and limitations of this research to support the author's learning for later research endeavours.

4.5 Challenges and Limitations of this Research Design

The interpretive research approach with semi-structured interviews and a focus group discussion has provided rich qualitative data. Nevertheless, this research design also had its limitations that are introduced following. For the author, an engineering and business graduate, and within his professional journey, holding a senior management position in a supply chain function with a strong focus on numerical figures, a positivist viewpoint would have been typical for this research, which would be in the natural scientist's tradition according to Saunders et al. (2009) or Bryman and Bell (2015). Due to his lack of experience with qualitative research, the author attached immense importance to thorough transcription to be able to use the data from the interviews for the following analysis. This lack of experience led to using an automated transcription tool to optimise the time-consuming transcription process, which strongly depended on clear pronunciation and the rhetoric of the interview partners. One of the interviews conducted over the telephone led to enormous difficulties in understanding the interview partner acoustically. Hence, the audio recording alone had to be re-listened several times to reproduce the interviewee's verbatim statements accurately.

Most of the semi-structured interviews were conducted in German, and only one of the semi-structured interviews was conducted in English. However, the German-speaking interview partners used strong local accents from southern Germany, Austria and Switzerland. To avoid losing the meaning of the interview partners' original narrative during translation, the author focused on most meaningful translating of the "original voice" of the interview partners. Besides the difficulties of correctly understanding the interview partners' local terms, accessing all information as initially planned was not possible. To enrich the information received during the semi-structured interviews that collected primary qualitative information, the author had planned to also gather secondary information from organisational documents, Supply Chain Risk Management policies and process documents, as discussed by Saunders et al. (2009). Once access had been given, triangulating multiple data sources would have been possible by combining documentary, short for *"documentary secondary data"* (Saunders et al. 2009, p. 258), with the semi-structured interview and would have aimed at gaining more information-rich data. During the pilot phase of this research project, it became evident that documentary was not accessible as it contained confidential information; some of the sought-after information did not exist at all due to the infancy of the Supply Chain Risk Management processes at the studied Hidden Champions.

Another challenge of this qualitative research project was the enormous amount of data collected, as envisioned by Gioia et al. (2013); this project revealed various risk factors identified in discussion with the interview partners which could not be used in this research project, while separated from the data relevant to this project.

The critical challenge of this research design, and its success based on the personal interaction between the study participants and the author, was the severe impact of the Covid-19 pandemic. For example, feedback to the partners from the semi-structured interviews was not possible. Also, piloting and implementing of the focus group discussion were obstructed by the Covid-19 pandemic. The author initially intended the focus group discussion to be conducted as a face-to-face meeting. Due to the online meeting format, which can be compared to a telephone interview, the analysis was more difficult, lacking the perception of body language. From the previous semi-structured interviews, non-verbal communication was one of the indicators that strengthened or weakened many of the interview partners' verbatim statements. To compensate for this lack of non-verbal communication, the author took notes wherever he realised unique behaviour from the focus group participants, such as variations in the discussion tone, and included that in the transcripts to emphasise the participants' narrative.

One of the substantial limitations of this research project is the contextual closeness between the interview partners and the author due to his professional background. According to Maier and Monahan (2010), this closeness leads to accusations of lacking distance between the researcher and the research subject, resulting in a lack of objectivity on the part of the researcher, both of which are necessary for valid interpretive research according to Saunders et al. (2009). In contrast, Brannick and Coghlan (2007) argue that through the process of conscious reflection, the information received can be critically questioned by the researcher and thus validated. Hence, the author applied a critical perspective using probing questions to discuss specific issues from a distinct perspective to validate the given information's plausibility.

However, the contextual closeness was also an advantage given the author's professional experience and seniority, enabling a more open interaction with the interview partners, thereby gaining the trust of the interview partners, which led to receiving a very rich narrative. On the other hand, this closeness and experience also had to be implemented with great care so that the author was not perceived as arrogant or overbearing, compromising the trust and the open interaction with the interview partners (Maier and Monahan 2010).

4.6 Chapter Summary

This chapter, Chapter 4, introduced the conceptual framework to research applied in this thesis. An interpretive research design has supported qualitative data collection through semi-structured interviews with senior managers responsible for Supply Chain Risk Management, located mainly in purchasing or supply chain management, to generate new concepts in Supply Chain Risk Management research. While entering into research with human participants, the ethical requirements for all phases of this research were presented to achieve the highest possible level of transparency. It discussed the pilot phase of this research project based on the literature review findings, probing the interview questions, and refining the author's qualitative research skills and competencies. The concept of data collection, coding and utilisation of the information received has proven qualitative rigour, delivering plausible and defensible conclusions. The knowledge gained during this research project and the presented data analysis and association with theoretical concepts identified would otherwise not have been possible. The qualitative data analysis findings were validated using focus group discussion; the research process, coding and analysis of the qualitative data, were probed using regular debriefing sessions with the research supervisors and peer review, providing additional criticality. The next chapter presents the results of the data analysis, substantiated with comprehensive references to the interview partners' narratives. The lively report on the interorganisational cooperation of Hidden Champions and their supply chain partners to manage and mitigate disruptive events in the upstream supply chain is organised based on the underlying data structure. It thus shows the development from raw data to aggregated dimensions.

Managing Risks in the Hidden Champions' Supply Chains

5

Chapter 5 presents the detailed results of the systematic data analysis based on Gioia et al. (2013) introduced in the previous chapter. An interpretive research approach allowed the author to use different qualitative research tools to understand and explain the Supply Chain Risk Management approaches between the Hidden Champions and their supply chain partners. This chapter provides the reader with a comprehensive picture of the research project that includes the results of the data analysis from the main phase of this research. It includes the semi-structured interviews with senior managers responsible for Supply Chain Risk Management, located mainly in purchasing or supply chain management and the focus group discussion to validate the findings of the semi-structured interviews in the main phase of this research. Based on the interview guideline, inspired by the expert interview to validate the findings from the literature review, and used to conduct the semi-structured interviews, this chapter presents and discusses the findings of these interviews. These findings are arranged according to the data structure resulting from the data analysis and thus reflect the logical structure that leads to cooperative Supply Chain Risk Management; the data analysis findings are grounded in existing theoretical models. Building on the translated narratives of the interview partners, the findings of this study are presented as an informative record providing a comprehensive picture of Supply Chain Risk Management in the interview partners' organisations. These findings are constantly compared with current research in the individual subject areas to create a comprehensive picture of the identified subject areas, which later serves to present the theoretical concept. Following the findings from the semi-structured

---

Supplementary Information The online version contains supplementary material available at https://doi.org/10.1007/978-3-658-50453-3_5.

© The Author(s), under exclusive license to Springer Fachmedien Wiesbaden GmbH, part of Springer Nature 2026

M. Mayer, *Spanning Organisational and Geographical Boundaries*,
https://doi.org/10.1007/978-3-658-50453-3_5

interviews, the author discusses the findings from the focus group discussion that served to validate the earlier findings from the semi-structured interviews.

5.1 Findings and Discussion from the Semi-structured Interviews

This section introduces and discusses the findings of the semi-structured interviews from the main phase of this study by grounding the findings of the data analysis in existing theoretical models. The contents of this report are developed around the quotes of the interview partners and present, on the one hand, the development of the introduced concept and discuss the theoretical backgrounds. On the other hand, according to Gioia et al. (2013), the quality of his research is substantiated based on the evidence provided by the interview partners. The extensive interview partners' narrative that emerged from the qualitative data collection on diverse topics convincingly contributes to presenting this study's results. To focus and thus limit the scope of this study, only the most convincing and self-explaining narratives of the interview partners (Ramachandran and Blakeslee 1998), are at the centre of this report. This discussion is arranged according to the data structure resulting from the preceding data analysis and thus reflects the logical structure leading to a cooperative Supply Chain Risk Management. This data structure identifies the thematic areas of Supply Chain Risk Management Processes, Supply Chain Risk Management Culture and Supply Chain Risk Management Cooperation required for a cooperative Supply Chain Risk Management to span organisational boundaries.

5.2 Supply Chain Risk Management Process

According to Zsidisin (2019) Supply Chain Risk Management has received growing attention in recent years. Due to the internationalisation of the Hidden Champions, there is a need to define and document Supply Chain Risk Management processes which became evident in the companies that participated in this study. Missing or informal processes discussed by Ritchie and Brindley (2000) or Ellegard (2008), along with a lacking definition of Supply Chain Risk Management as stated by Kilubi and Haasis (2015), posed challenges to the Hidden Champions and other supply chain members. It was therefore unclear what risk management in the supply chain included and how processes were to be carried out, or which steps must be carried out within the framework of Supply Chain Risk Management, as identified by Friday et al. (2018)

For the large part of interview partners, the Supply Chain Risk Management processes were defined or formalised in a way that the interview partners themselves saw need for improvement, voiced by Britzelmaier et al. (2015). The pragmatic use of spreadsheet tools which were commonly used in the working environment helped to facilitate the practical application of the Supply Chain Risk Management processes. Access to the documented results of the Supply Chain Risk Management process, however, was not possible. The example of Mill emphasises the mentioned limitation to access the documented results:

> *"I used to have a presentation on Risk Management, but I don't have it here now, maybe we are a bit too shirt-sleeved about it, but I always ask how much effort is involved, how much benefit is there?" (Mill 90).*

Hence, among the interview partners a clear understanding of the lack of a formalised Supply Chain Risk Management process was given; the instruments required for a Supply Chain Risk Management process were not yet available at GLIDE: *"But as I said, no instrumentalised instruments."* (Glide 19).

5.2.1 Supply Chain Risk Management Process Ownership

Hidden Champions are historically known for their high degree of backward vertical integration, defined by Simon as the percentage of *"total manufacturing done in-house"* (2009, p. 238). This high percentage of in-house manufacturing results in a low supply chain complexity, leading to Supply Chain Risk Management anchored in the purchasing organisation, as identified by Villena and Gioia (2018), or in the quality management organisation as identified by Chiarini (2017). However, during the semi-structured interviews, the author rarely identified Supply Chain Risk Management performed in the purchasing or supply chain management functions, and only one interview partner, Open, introduced a dedicated risk management organisation that stood out from the other Supply Chain Risk Management approaches. Through close cross-functional cooperation with the purchasing, operations, and quality department, the introduced Supply Chain Risk Management approach at OPEN indicated an existing intraorganisational cooperation, requesting to:

> *"Coordinate the risk situation in advance with purchasing, procurement and quality and then go there to carry out the risk audit."* (Open 39).

The high degree of vertical integration raised above served the Hidden Champions well as long local structures with linear upstream supply chains dominated.

5.2.2 Supply Chains with limited Complexity

Hidden Champions experienced limited complexity in their supply chains with a manageable number of supply chain partners, most of them within local or regional reach, as very graphically introduced by one interview partner: *"We are* [geographically] *very church tower related with our suppliers."* (Experiment 7).

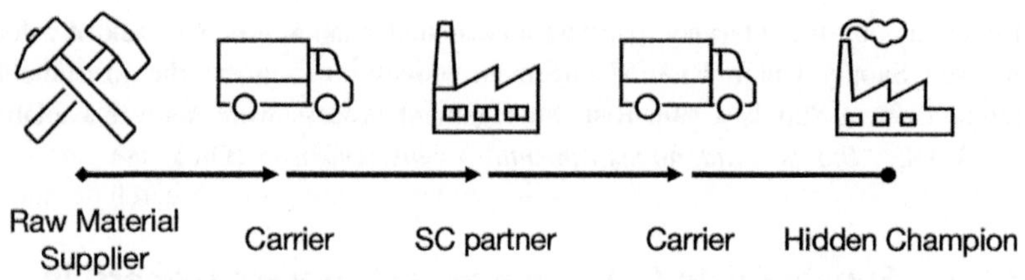

Fig. 5.1 Linear Upstream Supply Chain

Figure 5.1 exemplarily illustrates the limited complexity of a linear upstream supply chain with the Hidden Champions owning a position close to raw material suppliers and their supply chain partners, with the finished goods manufactured by the Hidden Champions. In contrast to the local structures and supply chains with limited complexity are Hidden Champions, following the globalisation approaches of their customers, as discussed by Audretsch et al. (2018).

5.2.3 Complex Upstream Supply Chains

These Hidden Champions experience increasing supply chain complexity with a purchasing or supply chain management organisation traditionally oriented towards the local market and employees neither prepared nor trained, nor knowing how to manage the challenges of this level of globalisation, identified by

Güller and Henke (2019). Conche reported of an incident where a supply chain partner in Asia could not produce urgently needed components, but

> "*... we had to fly in [components] from Switzerland so that we could supply our customers with* [...] *machines.*" (Conche 123).

The trend toward globalisation and the resulting increase in the number of overseas supply chain partners leads to complex supply-and-demand networks. The complexity of moving from the traditional linear upstream supply chain to a complex upstream supply chain or supply network (Villena 2019) is illustrated in Fig. 5.2.

Furthermore, the lack of adequate training for employees to deal with this complexity is one possible reason for the increased vulnerability of the Hidden Champions' supply chain discussed by Christopher and Lee (2004) and Jüttner (2005).

Further vulnerability of the upstream supply chain was induced by recent inorganic growth following the acquisition of new companies through the Hidden Champions studied, presented by one interview partner:

> "*You probably might have also done some research into how TRAVEL is positioned in the world and through acquisition we have gained new companies.*" (Travel 2).

This inorganic growth resulted in increased supply chain complexity due to a multitude of new supply chain partners as well as a heterogeneous corporate culture and intraorganisational language barriers. The lean philosophy of the supply chain at one of the Hidden Champions studied, who reported a five-days range of their production stock, "*... it is also shown how long the range is, 5 days as a rule.*" (Irrigate 119), is the reason for additional vulnerability, discussed by Norrman and Jansson (2004) or Ivanov et al. (2014). On the one hand, low inventories support the reduction of the overall supply chain cost (Farahani et al. 2014) and thus have a positive effect on the company's working capital (Zhao and Huchzermeier 2018; Tate et al. 2019); on the other hand, Christopher et al. (2003) or Ivanov (2021) state that low inventories pose a risk of interrupting the downstream supply chain in case of a disruptive event in the upstream supply chain.

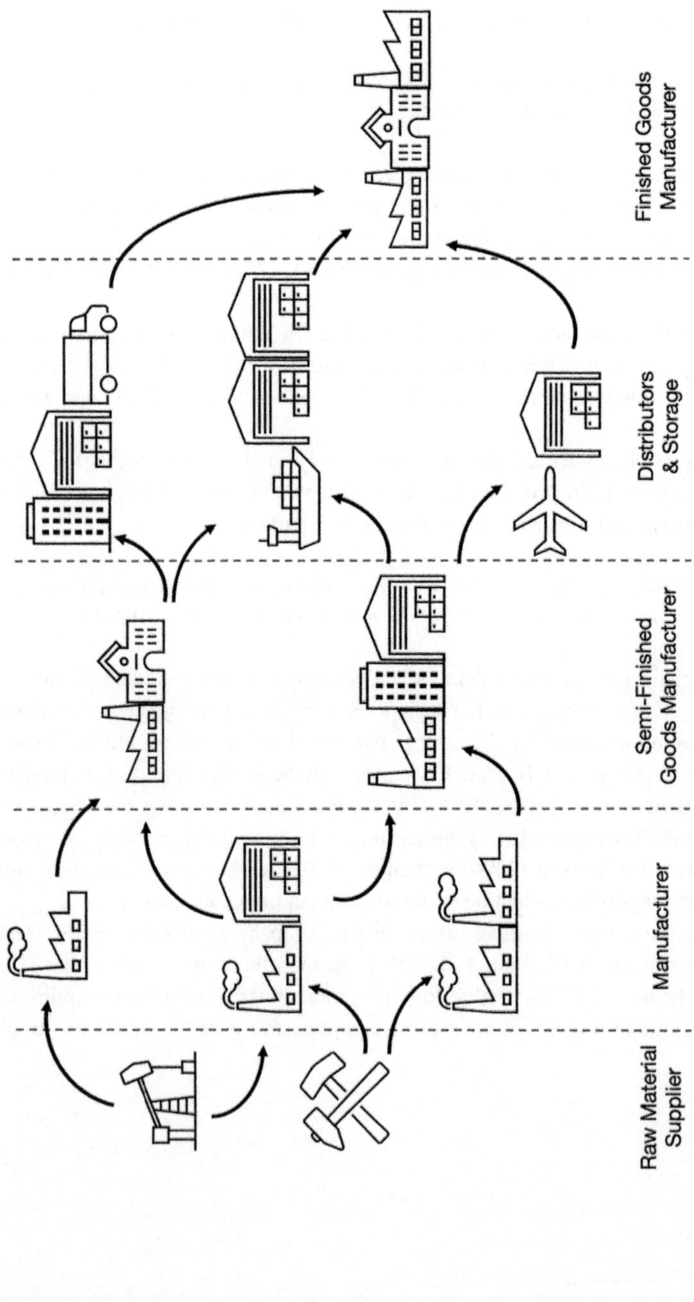

Fig. 5.2 Complex Upstream Supply Chain

5.2.4 Supply Chain Risk Management Designation

During the data analysis, it became evident that for many Hidden Champions, there was no designation, or a definition of Supply Chain Risk Management, helping to provide a mutual understanding and motivating all supply chain members to react quickly and with the appropriate measures in the event of a supply chain disruption as discussed by Diehl and Spinler (2013) or Sarker (2019). The following presents the responses from the individual interviews and the ambiguous designations of their Supply Chain Risk Management processes. For some interview partners, a designation for Supply Chain Risk Management did not exist, as Evacuates' answer to the question about a designation in the company demonstrates: "*Good question...* [pause]" (Evacuate 1). The ambiguity of terminology was also evident at TRAVEL, where it was not clear from the interview what Supply Chain Risk Management was called in the company:

> "*Good question, when everything flies apart, what is the name of the process that has been initiated. Well, we always call it Task Force.*" (Travel 169).

These findings are in line with the research of Ritchie and Brindley (2000), Ellegard (2008) or Britzelmaier et al. (2015) that discuss the formalisation of the Supply Chain Risk Management practices as with many of the interview partners no Supply Chain Risk Management designation was given, nor a process description was existing as introduced by Glide: "*We haven't instrumentalised it in that form yet.*" (Glide 4). Also LIFT did not have a separate name for the Supply Chain Risk Management process; it was described as Risk Management: "*We don't have a separate name for it, ...*" (Lift 1). For others again, the lack of designation may also result in the use of anglicisms, which one would like to avoid at MILL as with many other Hidden Champions in the DACH region:

> "*We don't speak any English here. We speak Swabian here, so, and with us it's just a normal supply chain, supply process.*" (Mill 11).

Other interview partners had a designation for Supply Chain Risk Management in their organisation, for example Experiment: "*We call it the same, supply chain risk management.*" (Experiment 156). This designation had its focus on managing risks in the upstream supply chain as confirmed by Conche:

> "*So this is called Supply Chain Risk Management, which is our official name.*" (Conche 30).

For one of the interview partners, the focus was on managing production-related risks, which included internal as well as external risks, as was the case with IRRIGATE:

> *"Risk Management: we do this once in production, that we naturally also do risk management:"* (Irrigate 29).

At OPEN, Supply Chain Risk Management was included in the higher-level enterprise risk management process and therefore bore this designation: *"Enterprise risk management"* (Open 2). The situation was comparable for CONTAIN, where risk management of the upstream supply chain was part of a higher-level process, the supplier relationship process:

> *"We have a higher-level process we call supplier relationships model. This supplier relationship model also includes the management of different risk factors"* (Contain 1).

5.2.5 Supply Chain Risk Management Process Definition

As with the designation of Supply Chain Risk Management, the definition of the Supply Chain Risk Management process also shows what the interview partners confirmed themselves. Supply Chain Risk Management processes were not clearly defined, corresponding process documentation was not available. This recurring theme also concerns Supply Chain Risk Management researchers (Manuj and Metzner 2008; Rao and Goldsby 2009; Diehl and Spinler 2013; König and Spinler 2016; Baryannis et al. 2019) while a missing definition hinders Supply Chain Risk Management from being implemented in Hidden Champions, state Diehl and Spinler (2013).

For those of the interview partners with standardised processes in place, the goal was to map complex processes to be as simple as possible in enterprise resource planning (ERP) systems and thus optimised the processes' performance, as stated by Irrigate following: *"I think our processes are simple. But everything is mapped in the* [ERP]*."* (Irrigate 305).

In contrast to the earlier narratives on informal Supply Chain Risk Management Processes, OPEN described a generic four-step risk management process. Within the first step (1) of this Supply Chain Risk Management process, a risk analysis took place to categorise the risks in the supply chain:

> *"We have the level 1, catalogue, risk categorisation of the different designations."* (Open 12).

In the second step (2), risks were evaluated by multiplying the probability of occurrence and the effect of the risk on the business:

> "In the second step, we have the step where we evaluate the risks and say: '*We have the probability of occurrence on the one hand and the effect of the risk on the other hand.*'" (Open 13).

In the third step (3), Open defined the risk management measures:

> *"A third step in the measures, in the risk-reducing measures, as we have the traditional: avoid; duplicate; separate."* (Open 14).

The process concluded with the fourth step (4), reviewing previously defined measures and implementing necessary amendments:

> *"In the fourth step evaluate all the risks again, we always have the periodic reviews for which we look at the risks of the measure, what is periodic, but what is, what do we do if risk changes fundamentally."* (Open 15).

This risk management process was to be applied throughout OPEN to create transparency by evaluating the possible effects of the risks in a comparable monetary form as one of the major aims of OPEN's risk management organisation.

As a result of the Supply Chain Risk Management process, which was an intraorganisational cooperation between the functional departments Risk Management, Quality Management and Purchasing, OPEN's critical partners and their potential impact on the business development were identified and included in a supplier database:

> *"What is very important is that we apply the system in the same way everywhere in order to get a reasonable comparability at the end of the day, as long as it is easy to understand."* (Open 17).

An approach that was intended to be comprehensive throughout OPEN's entire organisation.

5.2.6 Supply Chain Risk Management Approach

Within this thesis Supply Chain Risk Management is considered a proactive process to early identify upstream supply chain risks to mediate their management and mitigation. The findings of this research project suggest that Supply Chain Risk Management is merely a reaction to supply chain events, confirmed in recent research by Zsidisin and Henke (2019). While most of the interview partners did not perform risk analyses, proactive identification of risks in the upstream supply chain was impossible; therefore, cooperation with supply chain partners, which would help to manage and mitigate disruptive events early, is missing. Once disruptive events occur in the supply chain, the strategies for managing these events were defined ad hoc. As an example serves the narrative of Evacuate, who faced a situation where it was not possible to reach a supply chain partner for critical components for several weeks:

> "*We had a* [supply chain partner] *who moved and didn't get a new telecom connection there, and the production was hopelessly overloaded, that was last year when a lot of orders came in, and you couldn't reach him for weeks. 'Ok, now let's get in the car and drive there and see if the people are still there?' And these were essential components for our devices.*" (Evacuate 22).

In contrast to the reactive Supply Chain Risk Management approach presented above, other Hidden Champions in this study have started to develop their Supply Chain Risk Management organisations. Although they applied a first-hand approach to Supply Chain Risk Management, it contained more than only reactions to disruptive events in the upstream supply chain. At CONCHE risk analyses were performed to identify and assess risks and define appropriate measures to minimise risks:

> "*We produce a risk chart; this is a very simple multiplication, and the higher the value, the greater the focus we have to put on this case to minimise the risk.*" (Conche 72).

More proactive approaches could be identified by performing assessments before awarding business to supply chain partners. Comprehensive information on the supply chain partner, including reports from on-site visits and audits, was required before concluding a frame contract, to exclude possible risks explained Irrigate:

> "*In advance* [to the frame contract], *the risks, check, any information, supplier visits, corresponding qualification measures, auditing.*" (Irrigate 20).

Others described their Supply Chain Risk Management approaches performed through a task force, a cross-functional team established to manage and mitigate the disruptive event. Norrman and Jansson (2004) identify a task force as an emergency team constantly trained to cope with the existing Supply Chain Risk Management processes, reducing the consequences of disruptive events in the supply chain. Helmold (2014) adds that task forces are cross-functional teams with clear responsibilities supported by experienced process consultants. The concept of task forces was repeatedly introduced by the research participants where everyone would know the drills to follow when a disruption in the upstream supply chain occurred; the training provided to the members of the task force remains unclear:

> *"It must be said, no matter where things are on fire, we always make a task force. It's an established practice where everyone knows what's at stake: 'Now the roof is on fire.'"* (Travel 171).

Supply Chain Risk Management scholars such as Norrman and Jansson 2004; Henke et al. (2010 cited by Zsidisin and Henke 2019) and Schlüter (2019) support proactive approaches to enable cooperative Supply Chain Risk Management, which increases the resilience of the upstream supply chain in contrast to the earlier findings on reactive Supply Chain Risk Management approaches.

5.2.7 Supply Chain Risk Management Performance Indicators

During the discussion of Supply Chain Risk Management processes, the author studied the existence of key performance indicators (KPI) to assess the performance of Supply Chain Risk Management, as discussed by Manuj and Mentzer (2008). The operational performance of supply chain partners, represented by corresponding performance indicators such as OTIF (McLean 2017) for delivery performance and parts per million (PPM) for quality performance, discussed by Zsidisin et al. (2016) and Sarker (2019), was identified during the semi-structured interviews.

As with the designation and definition of Supply Chain Risk Management processes, the definition of key performance indicators mainly depends on the skills and experience of the owner-managers of the Hidden Champions. Once key performance indicators are seen as an essential part of the business management of the Hidden Champions, they find entrance in areas such as supply chain management, states Schlüter (2018) in his research on performance management in Hidden Champions. One of the challenges in the companies interviewed was that distinct indicators for Supply Chain Risk Management performance were not applied to support the Hidden Champions to measure or visualise Supply Chain Risk Management performance and achievements as identified by Mentzer and Firman (1994). Many respondents saw this as one of the significant difficulties of Supply Chain Risk Management expressed in the statement "*What you can't measure, you can't manage.*", which is attributed to either one of the management thinkers W. Edward Deming or Peter Drucker (Romeike and Stallinger 2021, p. 235). This in turn leads to the topic only gaining little intraorganisational acceptance, as stated by Conche:

> "*What I cannot measure and visualise, it is quite difficult to be able to set a focus there, even in discussions with the board.*" (Conche 118).

A comparable situation was expressed by Open, stating that the performance of Supply Chain Risk Management processes could only be reported through the lack of any disruptive events:

> "*The only way I can really objectively judge the performance of a risk management process is if nothing happens.*" (Open 80).

This narrative indicated the lack of indicators in use; although these would provide information about the performance of the Supply Chain Risk Management organisation and thus increase its transparency:

> "*Do good and report about it. We need our own public relations department for the Risk Management to tell the management: 'They work, they do a good job.'*" (Open 81).

Other interview partners had indicators in use to monitor the operational performance, discussed by Sarker (2019); measuring the current level of deliveries, such as quantities, the level of quality and the commercial performance of the supply chain partners, introduced as:

> "*KPIs on the plants' security of supply, i.e., delivery reliability, volumes, competitiveness—all this is ensured via KPIs.*" (Lift 16).

However, during the interviews, the author could not identify additional indicators to monitor the supply chain's increased resilience, as Cardoso et al. (2015) recommend, by applying Supply Chain Risk Management.

5.2.8 Supply Chain Risk Management Process Digitalisation

A topic that emerged during the interviews was the digitisation of Supply Chain Risk Management processes. The interview partners portrayed an openness to digital business process solutions managed via enterprise resource planning systems to monitor the supply chain and avoid additional maintenance work for their employees, which was expressed following by Cooper:

> *"It would be ideal to have kind of a system that could help you with the risk management perspective so that would be easier to keep track of things."* (Cooper 46).

By integrating the supply chain partners into the material requirement planning, as proposed by Wiengarten et al. (2016), previously manual planning, procuring, and payment processes could be automated recommends Glide:

> *"The first step is about our order process, the order confirmation electronically, the invoice electronically, the forecast confirmation electronically."* (Glide 56).

Process automation leads to more transparency and consequently to more resilience of the Hidden Champions' supply chain towards operational risks such as delivery failures, as confirmed by Schlüter (2019).

Despite the earlier portrayed openness, most of the Hidden Champions studied viewed digital solutions for Supply Chain Risk Management critically. Simon (2020a, b; 2021) also recognised this critical view in his recent publications, where he defines the digitalisation of Hidden Champions as part of their success strategies and calls for it to be advanced vigorously by the Hidden Champions. The reason for this critical view of digital solutions is the asymmetrical positions between the Supply Chain Risk Management solutions providers and the Hidden Champions as their customers, portrayed by Chen et al. (2017), that plays a significant role. Most consultants promoting such systems were not optimally adapting to the customer group of Hidden Champions. The interview partners reported frequent enquiries from digital Supply Chain Risk Management solutions providers offering standardised solutions, which was perceived more as a nuisance than a benefit, expressed by Conche:

> *"... I'm constantly flooded with requests from consultants who would like to make me happy, where I have to say: 'I'd like to find out through my network which the right partners are.'"* (Conche 23).

A dedicated solution tailored to the specific needs of Hidden Champions as customers did not seem to exist. These standardised solutions seemed to be met with rejection since it is the Hidden Champions who, according to Simon (2009, 2020), specifically put their customers in focus. Thus, selecting suitable providers for digital Supply Chain Risk Management solutions would occur in conversation with partners of the cross-organisational business network instead of relying on statements from consultants.

During the interviews, the author also identified capacity constraints that limited the willingness to invest in such solutions, as Hidden Champions already experienced capacity constraints for an extended period (Simon 2009). In this context, Irrigate expressed concerns about the implementation and data maintenance of a digital Supply Chain Risk Management solution, which he said would be very time-consuming: *"But this is a huge amount of data."* (Irrigate 345). As for any other significant investment, Hidden Champions strongly weigh whether digital Supply Chain Risk Management solutions would bring added value to their companies which Glas and Kleemann (2016) support in their research on using digital technologies in supply chain management. The Hidden Champions questioned if their companies were too small for the benefits of digital Supply Chain Risk Management solutions to outweigh the effort and costs of such solutions:

> *"In a normal medium-sized family business, I don't know if this is not shooting at sparrows with cannons."* (Mill 40).

A further challenge hindering the implementation of digital Supply Chain Risk Management tools was the employee training required. Implementing complex digital Supply Chain Risk Management solutions in Hidden Champions requires a changed competence profile of the employees. Zsidisin and Ritchie envisioned this changed competence profile already in their seminal work (2009) on Supply Chain Risk Management, now being confirmed in recent studies by scholars such as Glas and Kleemann (2016); Bals et al. (2019); Hofmann and Staiger (2020) or Weiß and Kamm (2020). The last point that raised scepticism towards digital Supply Chain Risk Management solutions was the false sense of security created among employees, meaning that their hard-won Supply Chain Risk Management awareness would be lost. As a result, employees would rely on the digital Supply Chain Risk Management solution to manage and mitigate disruptive events rather than their critical thinking skills, which was portrayed in the following statement of Irrigate:

> *"... and doesn't come into my office: 'Look, there is a little red light burning. Now I have told you that there is a risk', then I say: 'And now, now we look at each other!' That won't get us anywhere."* (Irrigate 351).

The discussion with Irrigate induced that digital Supply Chain Risk Management solutions would tell employees what to do, who did not know what to do. To overcome this knowledge gap on how to master disruptive events in the upstream supply chain, the Hidden Champions employees require training to increase their Supply Chain Risk Management competency. The following section will present and discuss how the Hidden Champions participating in this research project managed employee training as part of a corporate Supply Chain Risk Management culture.

5.3 Supply Chain Risk Management Culture

According to Simon (2009), the real difference between whether an organisation performs well or not lies in their corporate culture. A common understanding of the supply chain risks for all employees in the organisation would be necessary to be able to anchor the approach of an interorganisational Supply Chain Risk Management in the corporate culture of the Hidden Champions. To achieve common understanding, Supply Chain Risk Management training would be required—once for the employees in the purchasing or supply chain functions themselves, across functional departments and for the supply chain partners, as Sheffi and Rice (2005) or Zsidisin and Ritchie (2009) request. By creating common values and standards for all employees to better manage and mitigate disruptive events, the Hidden Champions would support a cooperative Supply Chain Risk Management culture to reduce the risks in the upstream supply chain, which is commonly supported by Sheffi et al. (2003); Christopher and Peck (2004) and Liu et al. (2018). Ultimately, Hidden Champions would need the appropriate employees for these interorganisational Supply Chain Risk Management processes; though, only a few of the interviewed Hidden Champions seemed to have established a Supply Chain Risk Management culture in their companies, however:

> *"An alliance only works if you have the same values and a similar cultural understanding."* (Conche 86).

The main enabler for a Supply Chain Risk Management culture would be the top management support, requested by Christopher and Peck (2004); Güller and Henke (2019); Kumar and Anbanandam (2020) and Weiß and Kamm (2020). Following top management support, the Supply Chain Risk Management training of employees in all areas of the company would help them to familiarise with the processes and requirements for a cooperative Supply Chain Risk Management. Regular audits of the Supply Chain Risk Management processes in the upstream supply chain would help to increase both intraorganisational and interorganisational risk awareness by means of fast identification and mitigation of supply chain risks at the supply chain partners find Christopher and Peck (2004) or Kumar and Anbanandam (2020).

5.3.1 Supply Chain Risk Management Awareness

A cooperative Supply Chain Risk Management culture is underpinned by corresponding Supply Chain Risk Management awareness, supported by training initiatives, as identified by Kern et al. (2011). This awareness includes, on the one hand, knowing and consistently applying the Hidden Champions' Supply Chain Risk Management processes and, on the other hand, knowing the risk awareness of one's own supply chain partners, as advocated by Christopher and Peck (2004). The Hidden Champions in this study should thus develop a Supply Chain Risk Management awareness to better manage and mitigate disruptive events in their supply chain, illustrated by the following comment of Irrigate:

> *"What happens if a* [...] *machine breaks down? In about 20 minutes, you won't get any more beer or schnitzel today. There will be no more plates."* (Irrigate 83).

The companies in this study have initially established Supply Chain Risk Management awareness. However, firmer anchoring of this awareness in the knowledge of the employees internally and across Hidden Champions boundaries still needs to be promoted. Companies should thus learn to manage and mitigate upstream risks throughout the entire process chain. The critical consciousness for more vital intraorganisational networking, based on the previously mentioned cross-functional training, was still needed to be established at EXPERIMENT:

> *"Over this complete process chain, we have to learn to manage our risks, which come along with it."* (Experiment 160).

Given the regional focus within their upstream supply chains, the Hidden Champions did not require solid Supply Chain Risk Management awareness in the past. With the increasing globalisation of Hidden Champions, those need to adapt their perception to the extensive risks in their upstream supply chains, leading to a changed Supply Chain Risk Management approach. Accompanied by change management processes, supported by Ates and Bititci (2011), this shift in risk awareness helped Hidden Champions achieve improved management and mitigation of disruptive events in the upstream supply chain says Irrigate:

> *"But I prefer my* [Buyer] *to make sure that he has short delivery times, is constantly on top of things and, once in a while, picks up the phone...."* (Irrigate 350).

5.3.2 Supply Chain Risk Management Training

Supply Chain Risk Management training is a systematic process that enables the employees' learning and development by increasing their competency. Supply Chain Risk Management training helps employees to acquire the ability to become highly adaptive to change, and they are enabled to perform in times of uncertainty find Tracey and Smith-Doerflein (2001); Beardwell et al. (2007); Schulze and Bals (2020) or Weiß and Kamm (2020). Therefore, as Blackhurst et al. (2011) or Sheffi (2019) discuss, Supply Chain Risk Management training would help employees to increase their awareness for and understanding of the necessary steps of the Supply Chain Risk Management process or related topics in their company to apply them correctly in case of supply chain disruptions.

One of the Hidden Champions' success strategies is what Simon defined as "*culture of high-performance*" (2009, p. 267). The interviewed organisations and their supply chain partners would need appropriate training to establish this high-performance culture in Supply Chain Risk Management, which would provide employees with the necessary knowledge and ability for cross-functional teamwork, according to Hallikas et al. (2019). However, the following statement described that Supply Chain Risk Management training was a blind spot on the agenda of the interview partners:

> [Training system for the employees?] *"If I sell myself well now, I say yes there is, by doing this process once a year."* (Lift 23).

As the only one of the interview partners, Cooper offered a specific training programme to his employees based on the identified training need to better understand the failure mode and effect analysis (FMEA) process that identifies and reduces risks (Stamatis 2019) as part of the risk analysis process (Curkovic et al. 2013):

> *"For the people that do not have that type of [engineering] education. Then we have to know, we provide the pressure courses, which can be both internal and external."* (Cooper 14).

Within other interviews, this research identified training approaches only in contract law, negotiation training and compliance as training on the job or in the context of joint workshops at internal conferences. For this training on the job, the Hidden Champions relied on the experience of their many long-serving employees. Due to an increasing strategic orientation in purchasing and supply chain management responsible for Supply Chain Risk Management, the long-serving employees would also need an extended training programme in the future to master the challenges of the internationalisation of the upstream supply chain reports Conche:

> *"There is clearly an open flank that we have, and if I want to drive this transition from operational to strategic, then I need a training programme as an accompanying measure, i.e. a personnel development and personnel training programme."* (Conche 92).

Other interview partners saw training itself as part of the Supply Chain Risk Management process as identified during the discussion with Travel:

> [Is training risk management?] *"I would say negotiations: 'Yes.' Because if someone had to prepare for them seriously, they would stumble over what is wrong with the supplier."* (Travel 98).

If, for example, one prepared well for subsequent negotiations because he received negotiation training, one could have quickly identified whether there were latent risks in the supply chain of the supply chain partner with whom the negotiation is planned.

Cross-functional training is a further aspect of Supply Chain Risk Management training to support intraorganisational cooperation and networking between the functional departments as reported earlier (Open 39). The intraorganisational cooperation of purchasing, quality management, development, production, and sales in the event of a supply chain disruption would enable a faster and better response to such disruptive events express Kern et al. (2011). This need for stronger social orientation and teamwork is supported by Bals et al. (2019); Crane and Hartwell (2019); Hallikas et al. (2019) or Weiß and Kamm (2020), asserting that cross-functional training would foster learning and raise a more substantial commitment of all departments involved in cooperative Supply Chain Risk Management. During one interview, Lift discussed this commitment:

> *"What is important is that the people work together, and thus done early in the development process."* (Lift 80).

Participants in this study attached importance to holding cross-functional lessons learnt meetings at the end of a project to benefit from past experiences. At EXPERIMENT, the focus lay on:

> *"Simply eliminate risks at this point for the future, avoid them, do not do them again."* (Experiment 76).

Moreover, those past experiences improved future projects and better reaction to possible disruptive events in the Hidden Champions' upstream supply chain. Although there was an emphasis on cross-functional lessons learnt meetings, this analysis did not identify formalised cross-functional training approaches.

Among other Supply Chain Risk Management scholars (Fawcett and Magan 2002; Christopher and Peck 2004, Cao and Zhang 2013; Hallikas et al. 2019), the author supports the idea of cross-functional training being essential to establish intraorganisational awareness for cooperative Supply Chain Risk Management. Promoting intraorganisational cooperation to regularly assess risks and increase the resilience of the upstream supply chain would serve the best interests of all supply chain members, including the Hidden Champions' customers. To achieve this cross-functional approach to Supply Chain Risk Management, Hidden Champions require the appropriate employees willing to work together with the other departments and abandon traditional practices. However, this cannot be achieved immediately but requires a strong will to change within the organisation. Conche reported a case where the development department employees carried out purchasing activities:

> *"Unfortunately, we still have the situation that we have developers who also like to procure a bit, right!"* (Conche 49).

This so-called *Maverick Buying*, the "non-compliant purchasing behaviour" (Karjalainen et al. 2008, p. 245) throughout functional departments other than purchasing, contrasts with the cross-functional approach to Supply Chain Risk Management because it nominates suppliers that have not undergone a previous risk assessment. To achieve a cross-functional approach to Supply Chain Risk Management, the above-mentioned cross-functional training or related human resource management activities are required to identify appropriate employees with team skills supporting the Hidden Champions. The next section will introduce these identified Supply Chain Risk Management talent management activities.

5.3.3 Supply Chain Risk Management Talent Management

To achieve a Supply Chain Risk Management culture that enables a permanently efficient cooperative Supply Chain Risk Management, companies need well-trained talents or employees with sound Supply Chain Risk Management knowledge and clear leadership and communication skills. This set of skills in employees could be achieved through Supply Chain Risk Management talent management, aiming to develop and retain these motivated and well-trained employees. In addition, Lewis and Heckman (2006); Kern et al. (2011) or Schulze et al. (2019) request a strategy for identifying and attracting new employees to the company. Supply Chain Risk Management talent or employee management programmes build on a pronounced Supply Chain Risk Management culture. A cooperative Supply Chain Risk Management culture leading to increased supply chain performance is attractive to existing and potential new employees seeking business success and job security (Hauswald et al. 2016). Talent management in supply chain management research has gained increasing interest in the recent two decades. Scholars like Hohenstein et al. (2014) and Garaus et al. (2016) also discuss human resource practices in Hidden Champions regarding onboarding, development and retaining talent. However, designated talent management approaches were not identified within the scope of this study; the individual activities to onboard, develop or retain employees identified will be discussed in the following.

During the interviews, the author identified different approaches to onboarding or integrating new employees into the Hidden Champions' Supply Chain Risk Management organisations. Some of which were more informal, which may be the reason for the small number of applicants on the one hand and the subsequently lamented high employee turnover in the younger generations on the other. Experiment emphasised this situation by stating:

> *"In the young generation, the bond with a company is no longer so pronounced."* (Experiment 123).

According to Schiemann (2014) or Seemiller and Grace (2017), those new employees require different, more engaging, forms of talent management compared to those of more experienced employees, confirmed by Leisy and Pyron (2009) in their study on talent management in global organisations. Other onboarding approaches were supported by a clear plan for future employee needs, backed by appropriate introductory training concepts. At IRRIGATE such concepts commence with product training which led to a better understanding of the risks:

> *"On the customer side, every buyer goes at least once before he even works, goes to work on our machines, an entire day, in order to also get along there, again risk analysis and supply security."* (Irrigate 82).

For some of Hidden Champions studied, the development of their employees was of significant importance. Employees were prepared to take on new strategic tasks, which required strong team players with a high work ethic and strong intrinsic motivation, described by Conche:

> *"For me, in the end, procurement is people business. You need to have the right people. Employees who have the appropriate comprehension skills and flexibility, who are interested* [in the topic]." (Conche 131).

This concept is also proposed by Schweiger (2015) or Bals et al. (2019) based on appropriate qualification programmes required to increase the maturity of the purchasing and supply chain management function.

A dominant issue for many of the interview partners was the scarcity of skilled employees, one of the significant challenges for the companies in this study, as stressed by Experiment: *"We only have very few applicants."* (Experiment 124). Hence, retaining talent through development or finding adequate successors was implemented differently by the Hidden Champions participating in this study. Some of the Hidden Champions understood the importance of talent management for their company, especially in Supply Chain Risk Management, because they wanted to retain good, experienced employees. However, there were also examples where one first had to go through a specific learning curve to realise that with a fluctuation in Supply Chain Risk Management, not only good employees left the company. Furthermore, it was the knowledge, a key core competence, that was also lost: *"[knowledge] is naturally lost when people leave."* (Evacuate 20), which according to Baptista Nunes et al. (2006) represents a significant risk to knowledge-intensive businesses. Others had dedicated, long-standing employees. On the one hand, Lehrer and Schmid (2015) justified this long-standing employment in the often rural location of the Hidden Champions. Far more significant was the corporate culture, which Mill characterised by a strong sense of togetherness between the employees and the top management:

> *"... because they form a healthy environment. When the big crisis was in 2009, the [owner] stepped up to the microphone and said, 'No one will be fired here!'"* (Mill 107),

also supported by Felden et al. (2019) or Audretsch et al. (2018), stating that the Hidden Champions' owning families strategic focus lies, among other things, on the preservation of employment.

5.4 Supply Chain Risk Management Cooperation

Within this study the author defines Supply Chain Risk Management cooperation as the proactive interorganisational cooperation between the Hidden Champions and their supply chain partners to mutually manage and mitigate disruptive events in the upstream supply chain. Supply Chain Risk Management cooperation builds on sharing risk-related information and aligning and developing Supply Chain Risk Management processes along with the partners of the upstream supply chain.

So far, there is surprisingly little discussion about a cooperative Supply Chain Risk Management approach in the extant literature. The majority of recent publications on joint management and mitigation of disruptive events in the supply chain discuss Supply Chain Risk Management collaboration (Cao and Zhang 2013; Kache and Seuring 2014; Li et al. 2015; Friday et al. 2018; Al-Doori 2019); collaboration or integration, both used interchangeably, refers to a close, contractually regulated connection between the supply chain members often centrally controlled via IT tools. However, Supply Chain Risk Management cooperation refers to the motivation for achieving a common goal, a cooperative Supply Chain Risk Management approach of autonomously acting supply chain members by sharing information and expertise. This approach presented, achieving common goals and aligning interests, according to Yan and Dooley (2013), is supported by Bals and Turkulainen (2021) in their research on buyer-supplier relationships in a global business landscape.

In this framework of Supply Chain Risk Management cooperation, the author discussed the influence of a close and long-term partnership between the Hidden Champions and their supply chain partners on interorganisational cooperation leading to trustful sharing of Supply Chain Risk Management information and agile communication between the members of the upstream supply chain. Further topics discussed that concerned Supply Chain Risk Management cooperation were the early supplier involvement (ESI) in new product development (NPD), the cooperation on Supply Chain Risk Management training and auditing and development of the supply chain partners' Supply Chain Risk Management processes, ultimately harmonising Supply Chain Risk Management strategy of the supply chain members.

The need for a cooperative Supply Chain Risk Management was clearly understood by most of the interview partners and was agreed to find entrance into the Supply Chain Risk Management approaches of the interview partners as confirmed by Conche:

> *"This is a good point that I take away from our discussion today. Risk management is absolutely an area that we can cooperate with suppliers."* (Conche 10).

5.4.1 Close Ties in the Market Segment

The close ties and good networking (Simon 2009) between the Hidden Champions and their supply chain partners who characteristically operate in a limited market in their respective industries are essential for Supply Chain Risk Management cooperation. This limited market also leads to more robust cooperation, not only from an economic perspective where open book calculations are made. It is furthermore a vital interest of mutual well-being that drives cooperation (Miller and Engemann 2019).

Often it was not the performance of the supply chain partners that decides in favour of the business relationship but the excellent and trustful relationship between the employees of the Hidden Champions and those of their supply chain partners. Experiment explains how this relationship led to successful cooperation in case of disruptive supply chain events, and

"... works very well when cooperation and partnership are really lived. We have been working with many suppliers for many years. This is a form of risk management that is difficult to replace by others." (Experiment 52).

Therefore, close cooperation on a human level and the long-term partnership is seen as part of a successful Supply Chain Risk Management in Hidden Champions.

5.4.2 Long-term Partnerships

The close and long-term partnerships between the Hidden Champions and their supply chain partners is considered one of the essential prerequisites for a successful cooperative Supply Chain Risk Management (Spekman et al. 2002; Chen et al. 2017). High-quality long-term partnerships based on mutual respect and principled-based behaviour are the framework for partnership cooperation identified by Wiedemann and Größler (2021), where supply chain members provide support in case of damage as confirmed by Cao and Zhang (2013). Hence, from a cooperative Supply Chain Risk Management perspective, it is vital to foster and maintain long-term relationships with supply chain partners, which is characteristic to Hidden Champions as illustrated by Mill:

"That's also a bit of the Swabian thing here; when you have family businesses, they are sometimes loyal to the death, if they can get along." (Mill 21).

The findings of this research project confirm a strong loyalty between Hidden Champions and their supply chain partners, encouraging long-term cooperation. This kind of relationship-inherent Supply Chain Risk Management is challenging to replace by other mechanisms. The mentality of the Hidden Champions in the DACH region was attributed to close ties and long-term relationships that enabled mutual support between Hidden Champion participants. Good relationships that lead to uninterrupted supply chains are more relevant for awarding new business than the price itself, states Glide:

"... let's call it relationship management with the supplier, that you get many parts earlier than other customers." (Glide 40).

However, uncooperative or opportunistic behaviour on the part of supply chain partners that abused trust could lead to phasing out the supply chain partner that resulted in termination of the business relationship find Cao and Zhang (2013) and commented by Mill as follows: *"They no longer eat bread with us."* (Mill 45).

5.4.3 Interorganisational Cooperation

One of the proven strengths of Hidden Champions is their proactive participation in the customer's development processes as discussed by Din et al. (2013). Other scholars like Parast et al. (2019) are consistent with the concept of joint development and innovation supporting the approach of interorganisational Supply Chain Risk Management cooperation to manage and mitigate disruptive supply chain events.

During the interviews, the concept of interorganisational Supply Chain Risk Management cooperation was addressed in additional characteristics. Among these characteristics were staff secondment discussed by Thomson and Arney (2015) and interorganisational support to cooperatively manage and mitigate disruptive supply chain events in new product developments projects. Travel described the interorganisational cooperation with a supply chain partner to manage and mitigate a recent supply chain disruption:

> *"This was put on the notice board everywhere: 'The company TRAVEL offers you elevenses and wishes you much energy and strength to rebuild the company.'"* (Travel 56).

Further interorganisational Supply Chain Risk Management cooperation was identified to reduce material cost fluctuation among the Hidden Champions and their supply chain partners, as well as applying instruments of *Supply Chain Finance* defined by Hofmann (2005). In their recent book, Tate et al. (2019) discuss those instruments to support supply chain partners when under financial stress to avoid disruptive events in the upstream supply chain. This cooperative approach to Supply Chain Risk Management aims to mediate the management and mitigation of disruptive supply chain events through interorganisational cooperation that, according to Parast et al. (2019), ensures the economic success and mutual benefit of cooperating supply chain members.

High material price fluctuation, especially in raw materials such as metals and plastic resins, represents a significant risk to the Hidden Champions' upstream supply chain contemplates Hofmann (2011). To project this raw material price fluctuation, CONCHE has internally developed a software application to project the raw material price fluctuation:

> *"In this case, we made projections. We have thought about what the effect on the business will be."* (Conche 81).

For others, not having such software developing capabilities, reducing material cost fluctuation in the upstream supply chain was one of the identified forms of interorganisational cooperation considered. Therefore, supply chain partners were allowed to source the raw material for products manufactured for the participating Hidden Champion from the same raw material source based on predefined raw material conditions, by offering the supply chain partner:

> *"'You call off based on our framework agreements, and we include this in our cost calculation.' So, for us, this is better cost or lower cost."* (Lift 53).

To mitigate the risk of raw material price fluctuation, the raw material quotas for the products to be procured along the upstream supply chain were identified with the supply chain partners during the planning phase and then ordered in bundled form from selected raw material suppliers known as *Supply Chain Sourcing* as discussed by Mohr (2009).

Financial support within the upstream supply chain is a concept employed by some of the Hidden Champions studied to support supply chain partners in financial distress. In most of the identified interviews this financial support provided ad hoc when one of the supply chain partners asked the Hidden Champion for support. It was mainly financing purchase orders, discussed by Zhao and Huchzermeier (2018) as down payments for new projects or inventory financing proposed by Vousinas (2019), providing financial support for the procurement of raw materials in the event of a liquidity shortage at a supply chain partner. When discussing the topic of financially supporting supply chain partners, Irrigate mentioned: "*By the time we realise it's the money, we'll be there.*" (Irrigate 180). Financial support in the form of supply chain finance (Hofmann 2005; Tate et al. 2019) was a topic that required formalisation for many of the interviewed Hidden Champions. Bals and Bals (2019) indicate that supply chain finance needs to receive more attention, both from academia and practitioners, with the conceptual development to be further advanced. These indications are consistent with this research's findings, where some of the Hidden Champions showed high interest in supply chain financing opportunities, while others were not using such opportunities or did not have defined strategies. Financial support was not seen as an altruistic approach; Hidden Champions aim to achieve shared benefits by applying a win-win strategy (Cooper and Ellram 1993) when financially supporting their supply chain partners. For supply chain partners requesting pre-payments or reduced payment terms, compensation for this financial support should have been agreed upon as explained by Travel:

> *"I always say: 'If you want something, I have to get something, right. Again and again, 'I pay you in five days, but then I want to keep the two per cent early payment discount.' For example, this can make quite a difference."* (Travel 279).

Hence, this cooperative approach would help both partners to be successful.

Staff secondment is another form of interorganisational cooperation, which is the compensation for the absence of staff at the supply chain partner by seconding specific staff members from the Hidden Champion to the supply chain partner over a brief period (Thomson and Arney 2015), as introduced by Travel:

> *"We had already received cries for help from our suppliers, they had a wave of flu or accidents, and the welders couldn't work. I can't get my material, and then we said, 'Okay, we'll give you two welders for two months, but they work on our projects.'"* (Travel 321).

The staff secondment helped the supply chain partners to successfully advance TRAVEL's projects to meet the promised delivery dates to the customer.

Sharing information among the upstream supply chain members is an important mechanism to gain shared understanding of Supply Chain Risk Management requirements as requested by Bowersox (1998 cited by Werner 2020); thus, to increase resilience or avoid inefficiencies and excess costs in the Hidden Champions' upstream supply chain introduced by Lee et al. (1997) in their seminal publication on the Bullwhip Effect. Information sharing was discussed extensively during the interviews of this research project. Proactive information sharing between upstream supply chain members increased resilience by early identification of disruptive events that impacted the effectiveness of the Hidden Champions' supply chain as promoted by Wieland and Wallenburg (2013) or required by Bals and Turkulainen (2019). Therefore, the Hidden Champions invested in supply chain partners that proactively shared information by awarding new business to them (Burkhardt 2019). One interview partner, Glide, mentioned:

> *"Suppliers that show such behaviour should be preferred for new orders."* (Glide 117).

However, many Hidden Champions allowed only for a limited amount of regular information sharing. Supplier days served as a suitable medium to encourage this regular information-sharing between the supply chain members, confirm Mayer et al. (2016). They served further to manage relations with supply chain partners and provided an opportunity to incentivise good performance agreed to by Beer et al. (2018). However, the operational workload of the daily business for some interview partners was taking its toll on strategic issues. It did not allow for such proactive ways of communication and relationship management like hosting a supplier day:

> *"Also, we want to do a supplier day, we always have, well, actual reason, I always say the daily business is rolling you over almost."* (Travel 242).

The early involvement of the supply chain partners can help to share the requirements for the cooperation and thus optimally utilise the advantages of the cooperation. Due to the high operational workload described above, the advantages of early involvement of supply chain partners in the product development process often cannot be realised. The lack of sharing the requirements, in turn, leads to the fact that the advantages of early involvement of the supply chain partners introduced following remain unused. Early supplier involvement, as discussed by Ragatz et al. (1997) and later Zsidisin and Smith (2005), is a proactive concept to optimise the supply cost during the early phase of the new product development process by cooperating with the supply chain partners in this early phase. Early involvement of the supply chain partners should be of enormous importance for the Hidden Champions, as it enables the social interaction of the supply chain members, getting to know their direct contacts in the relevant projects and the decision-makers of the cooperating organisations (Le Dain et al. 2020). It furthermore supports the identification of the strengths and capabilities of the supply chain partners and enables the exchange of requirements that the Hidden Champions have as a customer of their supply chain partners (Lau et al. 2010). This early cooperation in the development project thus increases the resilience of the upstream supply chain by identifying potential supply chain risks at an early stage, however requiring trustful communication examined by Zsidisin and Smith (2005). Most interview partners had supply chain partners who participated in the product development process early, allowing the Hidden Champions to outline requirements with the supply chain partners, which led to a significant reduction of disruptive events in the supply chain, as emphasised by Experiment:

> *"In my opinion, one way to improve it, is to involve the supplier as early as possible and take him on the journey so that he knows what is expected."* (Experiment 167).

However, some of the Hidden Champions interviewed were still in a learning process to integrate supply chain partners earlier in the product development process:

> *"Then I said: 'Yes, well, if you don't tell them how we want it.', then they throw everything [together], well packed though, but to identify ...."* (Travel 211).

Early involvement of supply chain partners would have fostered cooperation and information sharing to build mutual understanding upfront. Further on, early involvement of supply chain partners would result in increased quality and productivity by understanding the Hidden Champions requirements and thus helped avoid disruptive events in the upstream supply chain.

Electronic data interchange (EDI) is a standardised approach used by some of the Hidden Champions of this research project to share production forecasts with their key supply chain partners. Sharing production forecasts by integrating supply chain partners into such a joint planning process is defined as supply chain integration or supply chain collaboration by Cao and Zhang (2013). Both terms are used synonymously while describing the continuous interchange of information for joint planning of production processes among supply chain members to improve resilience and thus supply chain performance (Wieland and Wallenburg 2013). Joint planning and the interchange of production information was of high value to the participants of this study. Some of the participating Hidden Champions integrated their supply chain partners in a joint material requirement planning (MRP) via an EDI connection:

> *"There we have the EDI connection; there we have the whole automatic workflow in the area of payment processing."* (Conche 38).

The supply chain partners benefited from this integration by receiving clear expectations from the Hidden Champions, which enabled all supply chain members to secure supplies when following the explanations of Kulp et al. (2004). The interchange with the supply chain partners on joint planning created a high level of transparency in the material requirement planning, which provided further benefit for the participants of this study. Kache and Seuring (2014) highlight that this high level of transparency also offered a high degree of flexibility that enabled the Hidden Champions to react quickly to changes in the supply chain. This flexibility was important for Lift, as:

> *"That's what we learnt in 2009. It is in our mutual interest that we can ramp up and down the supply chain very quickly."* (Lift 64).

Despite the identified benefits of electronic data interchange, the interviews exposed that the participating Hidden Champions had not integrated all supply chain partners, which led to an increased planning effort. The interviews further revealed a lack of communication between the Hidden Champions and their supply chain partners on planning inconsistencies. Lift portrayed this lack of communication that could result in delivery delays or even disruptions in the supply chain:

> *"There is still a lot of potential for us there if we train the suppliers properly how these systems interact."* (Lift 68)

Increasing cooperation through information sharing using electronic data interchange with more significant numbers of supply chain partners and corresponding training on the potential of such systems, the Hidden Champions could have further increased the resilience of their upstream supply chain.

Agile communication is another aspect of sharing Supply Chain Risk Management information and is described by van Ruler (2019) as open and proactive communication; by omitting intermediate steps, agile communication addresses the decision-maker of supply chain partners directly and allows for facilitated information flows between the supply chain members assert Bals and Turkulainen (2021). Establishing agile communication helps to overcome the lack of communication between the Hidden Champions and their supply chain partners that the interviews revealed, leading to processes and contacts of the supply chain partners not being transparent. This non-transparency results in distorted information (Faisal et al. 2007) that leads to reactive Supply Chain Risk Management inhibiting the intended cooperation. Prolonged reaction times are a consequence, where swift and seamless communication could have triggered immediate reactions that could have prevented supply chain disruptions, also brought forward by Open:

> *"In a good partnership it is actually absolutely necessary that you also communicate with each other openly to a certain extent at an early stage."* (Open 118).

Especially during an upstream supply chain disruption, close and long-term partnership allowed for agile communication, which is a prerequisite to manage and mitigate supply chain disruption at the earliest possible occurrence. Travel agreed that agile communication that builds on trust between the supply chain members (Jayaram et al. 2014), is necessary in case of a disruptive supply chain event, that the communication chain's intermediate steps need to be left out and that there is direct communication channel between his organisation and the decision-makers of the affected supply chain partners:

> "I want a good relationship. I want to know the General Manager. That's right; it's actually part of the risk minimisation." (Travel 329).

The interviews further indicated that information on material requirement planning was the most frequent communication with supply chain partners for some of the Hidden Champions studied. In case of a disruption in the upstream supply chain, agile communication among the supply chain members was mandatory for fast management and mitigation of the disruptive event. Therefore, the author sought to understand how Hidden Champions organised the communication with their upstream supply chain partners in case of a disruption. During the discussion, Irrigate contributed to this:

> *"I have suppliers who know, when I contact them, then it's not about negotiating the price, but now it's burning somewhere in the supply chain."* (Irrigate 283).

Agile communication also includes the mutual exchange of critical information between the decision-makers of the supply chain members affected by the disruptive event. This agile communication enables quick decisions geared to the entire upstream supply chain to manage and mitigate the disruptive events as quickly as possible. For this type of agile communication, the contact details of the decision-makers at the supply chain partners are of high importance to the interview partners, so that direct communication between the Hidden Champions and the decision-makers at the supply chain partner is possible to prompt reactions and eliminate the supply chain disruption quickly similar to the findings of Norrman and Jansson (2004). A communication channel for all employees with supply chain risk-related information was established at MILL to allow for agile internal communication:

> *"They also talk about suppliers; we have an intranet, which means that we more or less have an* [Instant Messaging] *group where everyone is connected."* (Mill 85).

This communication channel allows all employees in the Supply Chain Risk Management organisation to alert the entire purchasing and supply chain management community if a disruption of the upstream supply chain called for coordinated risk management action (Sheffi 2019).

5.4.4 Interorganisational Supply Chain Risk Management Training

A further aspect of Supply Chain Risk Management cooperation is interorganisational training that extends the boundaries of the intraorganisational Supply Chain Risk Management training to the supply chain partners (Mohr and Sengupta 2002) in a cooperative form. This aspect is based on this study's theoretical framework, which seeks imitable and transferable practices spanning organisational boundaries to an interorganisational level as requested by Carter et al. (2017). Interorganisational Supply Chain Risk Management training shares the Hidden Champions' Supply Chain Risk Management requirements with the supply chain partners, assisting them to revise and update their Supply Chain Risk Management processes when needed (Bessant et al. 2003), which helps to mediate the management and mitigation of disruptive events resulting in increased overall supply chain performance (Manuj and Mentzer 2008). With interorganisational Supply Chain Risk Management training, the Hidden Champions share their experiences from previous projects in a lessons learnt workshop with their supply chain partners, as presented by Experiment during the interview:

> *"This is the ideal moment to have these conversations. You have to sit down with him* [supply chain partner]. *From the technical planning, the details, just everything that, communication and so on, didn't go well, you have to actually discuss."* (Experiment 79).

The lessons learnt workshop supported the Hidden Champions and their supply chain partners identify future improvement potentials (Güller and Henke 2019). The research project participants encouraged interorganisational training, especially in Supply Chain Risk Management, although training was mainly seen in continuous improvement processes and lean management workshops, as investigated in by Saier (2017), to support supply chain partners performing poorly. Like continuous improvement processes, relevant Supply Chain Risk Management topics could be analysed and improved with the supply chain partners as supported by Conche:

> *"There are certainly elements that can be addressed with the suppliers, for example, the topic of risk management, but I don't necessarily want to address contracting with the supplier."* (Conche 18).

While Conche supported interorganisational training on Supply Chain Risk Management, his rejection of interorganisational training in commercial topics resembles the findings of Mohr and Sengupta (2002) to prevent unwanted knowledge transfer. Other interview partners like Cooper hosted supplier days as a medium for interorganisational training approaches:

> *"We do annually what we call supplier day, in our three geographies where we operate, in the US and Europe and Asia. We do a supplier day, so we bring in the suppliers, we show them a little bit, we share with them our best practices, and our expectation, what we're doing."* (Cooper 18).

The awards presented at the supplier days were usually given a great level of attention in the corresponding market segments of the Hidden Champions express Beer et al. (2018). The related interest of the supply chain partners to participate in such supplier days offered opportunities for the Hidden Champions to promote interorganisational Supply Chain Risk Management training in cooperation with their supply chain partners. However, the scarcity of skilled employees (Simon 2009) at the interviewed Hidden Champions hindered the consistent employment of supplier days. Sharing experiences with supply chain partners, hence, the opportunities of interorganisational training largely remained unused (Hallikas et al. 2019). In summarising, interorganisational training requires more formalised approaches within the Hidden Champions studied. In some of the interviewed companies, interorganisational training only occurred after a disruptive event at the Hidden Champion to optimise the Total Cost of Ownership (Ellram 1993) in future projects due to the manifested risks.

5.4.5 Supply Chain Risk Management Process Audit

Process audits are often used in supplier management to analyse supplier processes in a methodically structured way, as introduced for example by Wieczorrek et al. (2019) to ensure compliance with previously defined standards (Helmold 2014). Auditing Supply Chain Risk Management processes according to Schröder (2017) is an option to evaluate the supply chain partners' Supply Chain Risk Management processes and to identify potential areas for improvement. If the supply chain partners already have established Supply Chain Risk Management processes in their companies, these findings can initiate best practice analyses to foster improvement of Supply Chain Risk Management processes among the supply chain members. Evaluation of Supply Chain Risk Management processes characteristically has a positive influence on the management and mitigation of supply chain disruptions and is seen as a result of interorganisational training state Hallikas and Lintukangas (2016). For this positive influence on the performance of Supply Chain Risk Management, evaluation is implicit, identifying the current state of these processes at the supply chain partners to identify potential improvement areas. Therefore, the consistent implementation of a regular auditing scheme for all supply chain partners and the definition of transparent evaluation criteria in a cross-functional approach are prerequisites.

For the Hidden Champions to understand the Supply Chain Risk Management processes of their supply chain partners, most of whom are also small and medium-sized enterprises and therefore prone to supply chain vulnerability (Villena and Gioia 2018), this research project sought to understand and explain how the Hidden Champions evaluated the Supply Chain Risk Management processes of their supply chain partners. During the interviews the author identified a lack of audits for the supply chain partners' Supply Chain Risk Management processes. In most cases, the suppliers of the supply chain partners were audited according to clearly defined criteria, with Supply Chain Risk Management not being a criterion, though. Many of the interview partners themselves identified this as an apparent deficit: "*... our suppliers are largely unaudited. This is a real weakness.*" (Experiment 46). Given the positive influences on the resilience of the upstream supply chain that auditing Supply Chain Risk Management processes entails, Hidden Champions should engage stronger in auditing their supply chain partners using such cooperative approaches, recommended by Christopher and Peck (2004), as they allowed for shared benefits, according to Cooper and Ellram (1993), which Open portrayed:

> "*This is also the case with our supplier audit. We always try to be cooperative and emphasise the win-win situation. So we say, 'We are all in the same boat, and if you are well, we are well.'*" (Open 128).

Contrary to the deficit in auditing supply chain partners identified above, interview partners were auditing before new business award to increase the understanding of the supply chain partners' Supply Chain Risk Management processes:

> *"At the same time, we let the supplier show us how he manages the risk assessment in his company."* (Conche 111).

This was supported by Irrigate, obtaining comprehensive information to exclude possible risks before releasing a frame contract to any supply chain partner:

> *"In the forefront, looking at the risks, any information, supplier visits, corresponding qualification measures, auditing."* (Irrigate 20).

The interview partners that described auditing the Supply Chain Risk Management process as a mandatory part of their cooperation with their supply chain partners supported the aim of this research project to understand and explain the mediating role of interorganisational cooperation requiring training and development of employees between Hidden Champions in the DACH region and their supply chain partners on efficient Supply Chain Risk Management.

5.4.6 Supply Chain Risk Management Development

The nucleus for cooperatively developing Supply Chain Risk Management processes is auditing these processes. These audits are the basis for further process development and joint improvement, resulting in process harmonisation between the supply chain members. The earlier findings of this research project indicate that potential for improvement would lie in cooperative activities such as joint Supply Chain Risk Management training approaches and a revision of the Supply Chain Risk Management processes to increase the engagement of supply chain partners in a cooperative Supply Chain Risk Management process as supported by Hallikas et al. (2019). The interviews indicate further that Supply Chain Risk Management development was a continuous and long-lasting process requiring sufficient skilled employees within the Hidden Champions' organisations, which was seen as critical by Cooper:

> *"We do want to collaborate with suppliers, but we have a huge responsibility on our hands, so we don't have such a huge manpower or resources to try to go and improve our supplies."* (Cooper 33).

Further interview partners confirmed that understanding Supply Chain Risk Management requirements is subject to intense cooperation, leading to a very time-consuming process as accentuated by Experiment: *"It takes a year to bring a supplier on track."* (Experiment 198). One interview partner, Open, however, mentioned that developing supply chain partners in a partnership way had improved the performance of the supply chain:

> *"We always try to work together as partners to bring about a better situation within the supply chain, don't we!"* (Open 46).

The interviews revealed that to support supply chain partners in becoming acquainted with and adapting the Hidden Champions' specific Supply Chain Risk Management requirements or processes which should be transferable between the supply chain members, a Supply Chain Risk Management development would be needed, which is also supported by Helmold (2014).

5.4.7 Supply Chain Risk Management Process Harmonisation

The harmonisation of the underlying processes is a further aspect of Supply Chain Risk Management cooperation in which the Hidden Champions seek to harmonise the relevant processes and strategies with their supply chain partners to achieve increased benefits for the entire upstream supply chain, seen as a benefit from cooperation by Cao and Zhang (2013). Just as the Hidden Champions align their targets with those of their customers (Wagner 2010), the Supply Chain Risk Management process harmonisation leads to a joint understanding, this joint understanding is supported by Bals and Turkulainen (2021) in their research on buyer-supplier relationships in a global business landscape. Therefore, harmonisation of Supply Chain Risk Management processes should be done cooperatively to create a shared understanding of the specific requirements. This shared understanding of specific Supply Chain Risk Management requirements enables the supply chain members to improve their Supply Chain Risk Management cooperation. Once different requirements exist, these are to be discussed and, if necessary, adapted to enable mutual learning. Conche saw this shared understanding to improve the cooperation between supply chain members:

> *"The better the common understanding is, the easier it is to work together on the issues where we really see risks. Good point."* (Conche 9).

In the context of this research project, the focus lies on harmonising the Supply Chain Risk Management processes of the individual supply chain members respecting individual specifications. This process harmonisation further supports a common understanding of the organisational culture, values, and expectations towards Supply Chain Risk Management once shared with the supply chain partners. Mutual respect helps to take a critical look at one's own Supply Chain Risk Management processes and possibly improves them rather than merely standardising existing procedures at supply chain partners (Friday et al. 2018).

5.4.8 Summary of the Findings from the Semi-structured Interviews

The findings from the semi-structured interviews disclose that the majority of the participating Hidden Champions had no, or only rudimentary Supply Chain Risk Management processes established, which had already become evident in the lack of a clear designation for Supply Chain Risk Management. Most interview partners confirmed that their companies did not have formalised Supply Chain Risk Management processes. Risk analyses were carried out and documented using spreadsheets. Approaches to a digitalisation of Supply Chain Risk Management processes were often discussed with digital solutions providers but perceived as too time-consuming to implement. In most of the participating Hidden Champions, Supply Chain Risk Management approaches were a reaction to events in the upstream supply chain, without clearly defined indicators to measure the performance of Supply Chain Risk Management. From the interviews, the author understood that some participants defined these indicators only after the Hidden Champion disruption occurred. Hence, a distinct Supply Chain Risk Management culture was in its initial stages at the time of the interviews. Many companies had in common that they work in, what they call, task forces to manage risks. However, it was unclear what these task forces were composed of, which functions were involved, or how the participants were trained.

With a focus on the framework of Supply Chain Risk Management Culture in the Hidden Champions, more training approaches are required to promote risk awareness among the employees in purchasing and supply chain management. A cooperative Supply Chain Risk Management approach requires cross-functional training to be established in the Hidden Champions to promote cross-functional risk awareness, resulting in intraorganisational cooperation to assess the risks regularly and thus increase the resilience of the upstream supply chain.

The Hidden Champions participating in this study reported having close cooperation with their supply chain partners that supported their cooperation as reported by Travel:

> "... for the largest part in the cooperation with the suppliers, we have a very good relationship." (Travel 327).

However, it was not clear to the Hidden Champions how their supply chain partners defined their Supply Chain Risk Management processes; hence, the development and coordination of Supply Chain Risk Management processes at the supply chain partners required further attention. The next section will introduce the findings and discussion of the focus group discussion that served to validate the findings of the semi-structured interviews discussed above.

5.5 Findings and Discussion from the Focus Group Discussion

Following the semi-structured interviews in the main phase of this study, a focus group discussion served to validate the findings of the earlier data analysis and to underpin this qualitative study's trustworthiness and verify whether the results of the data analysis were consistent with the respondents' comments displayed earlier. The focus group discussion participants were introduced to the aim of this research project, which was to understand and explain the mediating role of interorganisational cooperation requiring training and development of employees between the Hidden Champions participating in this research project and their supply chain partners. Furthermore, this study's research design, including identifying the Hidden Champions to participate, was introduced to increase trustworthiness. The information given to the participants of the focus group discussion ensured that the data analysis and the introduced model to cooperative Supply Chain Risk Management were plausible and made sense to those experiencing the phenomenon of Supply Chain Risk Management in their business life. With the focus group discussion, following the Covid-19 pandemic-induced lockdown in the three countries of the DACH region, the number of participants in the focus group discussion was smaller than expected. Some of the participants who had already confirmed their attendance were not available during the focus group discussion. Despite enquiries, reasons for their absence remained unclear. Just as important as absence or those directly participating during the focus group discussion were the cancellations received from interview partners who could not participate. They indicated that the Supply Chain Risk Management approaches in use at some absent participants did not support an early identification of potential risks that could threaten the viability (Ivanov 2021) of the supply chain:

> *"Dear Mr Mayer, unfortunately, I cannot attend this appointment. We are currently due to the KUA* [short-time work], *production stops, supply chain bottlenecks, ... We are very much in operational mode. Actually, this is exactly the situation that should be prevented by effective risk management. A crisis like this was perhaps to be expected, but no one is prepared for it. I would be happy to get back to you at a later date."* Glide Cancellation (2020).

The original version of this cancellation in German is included in Online Resource 5.

5.5.1 Cross-Sectoral Sampling Strategy

The focus group discussion opened with feedback on the research design itself. The cross-sectoral approach within the mechanical engineering industry of the DACH region was perceived as being highly beneficial by the interview partners as it avoids a sector-specific bias. It also shows the different approaches to Supply Chain Risk Management in the respective companies and thus enabled interorganisational learning, which resembled a positive experience that LIFT had in the weeks before the focus group discussion:

> *"I think it's a great thing and I felt the same way. In the last few weeks, first of all, it's been very exciting. In the last few weeks, I've talked to entrepreneurs and board members that I've never talked to in the last three years because they simply don't belong to my business sector."* (FGD 63 (Lift)).

5.5.2 Discussing Supply Chain Risk Management Processes

The cooperative Supply Chain Risk Management approach introduced was understood as a beneficial concept to the focus group discussion participants that was already established to some extent in the participants' organisations but still informal, Mill commented this:

> *"I think we all live the concept to some extent. But whether we really put it down on paper... ?"* (FGD 1 (MILL)).

Mill's statement confirms that the Supply Chain Risk Management processes still need appropriate definition and formalisation, as previously expressed by Ritchie and Brindley (2000); Ellegard (2008) or Britzelmaier et al. (2015). Mill further expressed that the decision-making processes were not sufficiently defined, i.e. whether the associated organisation had sufficient decision-making competencies. However, these decision-making competencies and the support of the company's top management are significant for quick decisions in the event of a supply chain disruption state Autry and Sanders (2009). Mill identified this top management support as critical lessons learnt from the early stages of the Covid-19 pandemic for the participating organisations. The need for a more vital awareness to support the Supply Chain Risk Management processes, also among the top management (Autry and Sanders 2009), was further debated during the focus group discussion:

> *"What I have now also learnt from the Corona crisis, or what we have also learnt, is the following. What is perhaps even more important is that you also establish who makes decisions, because they* [top management] *are not into risk management."* (FGD 54 (Mill)).

5.5.3 Unpronounced Supply Chain Risk Management Culture

Supply Chain Risk Management awareness is reason to foster the intraorganisational sharing of common values and standards that led to a vital Supply Chain Risk Management culture, which helped focus group participants defining a modus operandi that guided them on how risks are managed and mitigated, early at the beginning of the Covid-19 pandemic:

> *"... but what is very important, is what I saw in another company where I talked to someone. These discussions were not clear, and the mode was not clear, and these companies have been talking past each other for much longer."* (FGD 99 (Lift)).

For the supply chain to stay alive or viable even when hit by disruptive events as defined by Ivanov (2021), such a modus operandi needs to be found that is adaptive to the respective market segment or industry sector in which the Hidden Champions operate. Different foci prevailed in the various market segments the focus group participants operated in. One participant focused on liquidity; the other focused on securing supplies to make sure customers in the do-it-yourself (DIY) businesses could receive deliveries. Mill described the different foci in the following statement

> *"Due to short-time work, the DIY stores have gone through the roof.* [OMEGA] *is ordering like crazy. For us, liquidity was never a question. For us, the question was "We have to be able to deliver, because people are sitting at home working in the garden and renovating. That was very important for us. We had a three-month supply shipped in from our factory in Italy, and we did well with that because they were actually dead eight weeks now. I think, as you said, Mr* [LIFT]: *'You have to decide in which market you are operating, and what is important?' "* (FGD 101 (Mill)).

The semi-structured interviews revealed a pending Supply Chain Risk Management awareness. The Covid-19 pandemic has promoted this Supply Chain Risk Management awareness to understand the risks in the upstream supply chain that led to the need of an agile Supply Chain Risk Management organisation which supported flexibility to manage and mitigate disruptive events quickly. If Supply Chain Risk Management existed and was practised, one would be prepared to deal with uncertainty and to react quickly to disruptive supply chain events, as voiced by Lift:

> *"The last three major crises were not really on our radar, and we were always surprised. My learning from this is that you need risk management, you have to try to minimise risks. But you also need an organisation that can deal with challenges in an agile way and adapt very quickly. And you need this flexibility, this quick reaction, this agility. Of course, you have to try to minimise the obvious risks as much as possible."* (FGD 110 (Lift)).

The focus group discussion participants validated the findings of the semi-structured interviews that revealed the inexistence of cross-functional training; once existing, cross-functional training creates awareness for Supply Chain Risk Management when assembling a task force with members from different functional areas in the company to identify problems and define problem-solving approaches. For the task force members to better cooperate once managing and mitigating disruptions in the supply chain, as called for by Bals et al. (2019), cross-functional training is necessary to promote complementary competencies. Experiment explained the assignment of task forces in the following way:

> *"At least that's how it was with us when, in the course of Corona, we said: 'Hey, we need a task force here' that is made up of different functional team members, but of course also of members of our own supply chain area, where we sit down and say: 'So, what's the problem? What are the consequences? And how do you deal with it?'"* (FGD 51 (Experiment)).

5.5.4 Isolated Supply Chain Risk Management

The cooperative Supply Chain Risk Management approach based on the interview findings presented would help standardise and improve Supply Chain Risk Management processes that could be established in the Hidden Champions as part of a change management approach (Ates and Bititci 2011). The presented concept of Supply Chain Risk Management cooperation across organisational boundaries, grounded in process improvement based on the four-step plan-do-check-act (PDCA) cycle or Deming wheel (Saier 2017), was perceived as a good occasion to turn identified risks in the upstream supply chain into opportunities, as characterised by Elahi (2013). This discussion on business opportunities was requested by Lift:

> *"I would also like to talk more about opportunities in these discussions and see it that way. And when you show this PDCA, they should also go more into these issues of opportunities."* (FGD 31 (Lift)).

The long-term and stable partnerships forming close ties between the Hidden Champions and their supply chain partners surfaced during the focus group discussion, with the participants confirming the increasing supply chain resilience induced through the long-term and stable partnerships. According to Experiment, those Hidden Champions that had long-term and stable relationships with their supply chain partners had managed the supply chain disruptions well that rose from the Covid-19 pandemic:

> *"I can confirm that. Here, typically, purchasing was criticised for purchasing around the church tower. That was before the crisis. During the crisis, it turned out that this network was incredibly stable. We got through the crisis without any problems. As I said, there are disadvantages that come with it, that's quite clear. But it proved its worth, especially during the crisis. And against the background, yes, that is one learning."* (FGD 84 (Experiment)).

The supporting effects of long-term partnerships in disruptive events where supply chain partners are ready to help based on long-standing relationships with the focus group participants companies require an even stronger focus in the future. However, also disadvantages are considered from long-term relationships while having a limiting effect; on the one hand, reduced flexibility when considering changing supply chain partners as recognised by Stevenson and Spring (2007) and on the other, when considering the balance between partnership and dependence described by Chen et al. (2017); Bals and Turkulainen (2021) or Wiedemann and Größler (2021). This buyer-supplier relationship (BSR) was considered by Lift:

> *"So I also find that very exciting, and I'm convinced it's more about partnerships. But I don't yet know how to find the balance between partnership and dependence."* (FGD 70 (LIFT)).

Interorganisational Supply Chain Risk Management training helps to improve the Supply Chain Risk Management processes within the upstream supply chain especially for smaller partners state Villena and Gioia (2018), which also emphasised the smaller supply chain partners' missing control over tools and processes. An issue indicated by Lift:

> *"Because they might be smaller, but they are still important for us because even a small company can deliver a key component and I know that they don't have their processes and tools under control, so it certainly makes sense to train them and they're happy to accept it."* (FGD 60 (Lift)).

In discussions prior to this focus group discussion, one interview partner learnt a lot of additional information for his own company across sectors by avoiding an overconfident approach. Close ties and the good and long-term partnership on an equal level (Kähkönen 2014) within the upstream supply chain were especially important to the participants of the focus group discussion in times of crisis. A cooperative Supply Chain Risk Management approach within the upstream supply chain was widely supported, enabling mutual learning to help to improve Supply Chain Risk Management processes and risk assessment in one's own company based, on the experience of the supply chain partners itself:

> *"On the other hand, I am convinced that we have already learnt a great deal from suppliers, from large suppliers, about how to do our processes properly, about our risk assessment."* (FGD 61 (Lift)).

Interorganisational training approaches were supported by other interview partners although difficult to implement given challenges in day-to-day business and the existing cost pressure. Ultimately, the implementation of such approaches can only be successful within organisations and spanning across to supply chain partners depending on and are subject to the support and Supply Chain Risk Management awareness of the individual business owners (Autry and Sanders 2009). The experience shared by Mill enlightens this situation:

> *"That's when one tells the other 'You think my people need to be trained just because you think you have to do risk management'."* (FGD 11 (Mill)).

Financial support provided to supply chain partners was also a topic during the focus group discussion. With the onset of the Covid-19 pandemic, some of the Hidden Champions participating in the focus group discussion geared their corporate strategy towards a high level of liquidity. Therefore, the payment terms of the supply chain partners were adjusted accordingly, considering their financial situation, so that the Hidden Champions could sustain high liquidity during the crisis and to avoid potential challenges. By means of reverse factoring, discussed by Lekkakos and Serrano (2016) or Farris and Hightower (2019), LIFT made sure that their supply chain partners did not suffer any liquidity bottlenecks, and their invoices were paid before the due date:

> *"So we did everything like that and also looked at the suppliers to see where we could extend the payment terms, if that was realistic. But at the same time, we introduced a reverse factoring system a long time ago, where we pay the suppliers immediately via a bank, but they then use our financing costs for this, so to speak. And that has worked very well during the crisis. Because we get the money extremely cheaply, we actually have interim financing for the suppliers through this reverse factoring."* (FGD 106 (Lift)).

Sharing Supply Chain Risk Management information through intensive discussions throughout the supply chain enabled focus group participants to identify the current situation of their customers and supply chain partners from the beginning of the Covid-19 pandemic, which helped to increase the resilience of the supply chain as found by Wieland and Wallenburg (2013). Experiment following introduced this intensive communication that contributed to the subsequent measures to manage and mitigate the risks materialised in both the upstream and the downstream supply chains.

> *"That is to say, can our customers continue to maintain their construction sites or do they have to close them down. So, that we first knew whether we could continue to deliver on the sales side. Yes or no? And in the same breath we discussed intensively with our suppliers to find out the situation there, to establish what risks we have on this side? And in spite of all this, we have of course also increased our stocks and, against this background, have come through the crisis without any problems."* (FGD 107 (Experiment)).

Agile communication (Van Ruler 2019), the open, proactive and direct sharing of information between upstream supply chain members to increase resilience by early identification of disruptive events (Wieland and Wallenburg 2013; Bals and Turkulainen 2019), was a topic that provoked lively discussion among the focus group participants who were already successfully using agile communication for intraorganisational communication. Communication in the context of supply chain disruptions beyond the Hidden Champions' boundaries is supported and works well between family-run businesses; business owners often communicate directly. However, as companies grow in size, direct contact with those in charge became increasingly complex, as Mill laments:

> "The larger the corporation, the more difficult it is to find someone responsible for doing this." (FGD 8 (MILL)).

An information platform with supply chain partners is generally considered positive and should be integrated in Supply Chain Risk Management processes. Nevertheless, reservations about the implementation were still present, although the prompt exchange of information in the event of a supply chain disruption was supported, usually taking place at a more informal level. The background for Experiments' reservations was the exchange of information with the customer in the event of problems:

> *"But, each of us is not only a manufacturer who depends on suppliers, but also acts as a supplier to its customers. And they know only too well that typically the last person I inform when I have a problem is my customer. After all, I am trying to work out solutions to eliminate my risk."* (FGD 38 (EXPERIMENT)).

Thus, cooperation in partnership to solve disruptions does not seem to exist.

5.5.5 Summary of the Findings from the Focus Group Discussion

With the focus group discussion following the main phase of this study after six months, the interview partners were able to critically discuss the findings and the introduced concept for the continuous process improvement towards a cooperative Supply Chain Risk Management. Hence, the focus group discussion validated the findings of the semi-structured interviews of this research as well as it addressed the related research question and the chosen concept to interview Hidden Champions within the mechanical engineering industry in the cross-border area of the DACH region. The focus group discussion itself and the presented cooperative approach to improving Supply Chain Risk Management processes were evaluated as positive while establishing an information exchange to discuss the different approaches to manage and mitigate the risk in the supply chains of the participating companies amid the Covid-19 pandemic.

The close ties and the good and long-term partnership at an equal level within the upstream supply chain are significant to the focus group discussion participants in times of crisis while offering mutual support, which leads to resilient supply chains. However, these long-term partnerships are criticised for narrowing flexibility and increasing dependency on supply chain partners. Despite this criticism, the introduced cooperative Supply Chain Risk Management approach seems to find even more support in companies in the future while strengthening the partnership in the upstream supply chain, thus resulting in interorganisational cooperation that is essential to solving supply chain disruption ahead of time. The focus group participants confirm supporting the cooperation with their supply chain partners by instruments such as supply chain finance, which aims to improve the Hidden Champions' liquidity situation and that of their supply chain partners. During the focus group discussion, the participants emphasised the increasing exploitation of this cooperation that the Hidden Champions disliked. The goal here is to have a balanced power relationship with the supply chain partners.

Intraorganisational training can help further to establish shared values and standards and increase awareness of Supply Chain Risk Management requirements, which is vital to establish a Supply Chain Risk Management culture within the Hidden Champions that supports the resilience of the upstream supply chains. However, some focus group participants voiced concerns about the additional costs for the company due to an oversized Supply Chain Risk Management organisation. The focus group discussion confirmed that Hidden Champions with a high awareness are well prepared for possible disruptive events in their upstream supply chains and thus know how to manage such disruptive events and unforeseen events quickly and flexibly, which counters the above critical voices. In this way, companies can avoid additional costs due to production downtimes or even insolvencies. Although the participants of the focus group discussion supported training, none of the focus group discussion participants addressed Supply Chain Risk Management talent management including training and development of employees, although mentioned as one of the biggest challenges for the Hidden Champions during the semi-structured interviews.

Agile communication was a high priority for the focus group participants, particularly in companies with family businesses as supply chain partners where direct contact with the company owner was possible to manage and mitigate disruptive events. With large publicly owned corporations, this direct contact seemed impossible. Interorganisational training helps to improve Supply Chain Risk Management in smaller supply chain partners; however, Hidden Champions should be open to the benefits of interorganisational training instead of relying overconfidently on one's Supply Chain Risk Management capabilities. To create a mutual understanding of their Supply Chain Risk Management requirements, it is essential that the Hidden Champions clearly describe these requirements to their supply chain partners. Furthermore, there seem to be difficulties in the implementation of interorganisational training itself. On the one hand, the Hidden Champions need to address the topic of interorganisational training remarkably high up in the hierarchy of the supply chain partners; on the other hand, the corresponding capacities in the purchasing and supply chain management functions at the Hidden Champions have to be available. While the focus group participants claim to have already established the concept of cooperative Supply Chain Risk Management, albeit informally, they presented themselves as largely unaffected by the consequences of the Covid-19 pandemic. As a result of the focus group discussion and cancellation received, the author identified a solid need for establishing a cooperative approach to Supply Chain Risk Management that standardises or improves Supply Chain Risk Management processes to increase the supply chain performance.

5.6 Chapter Summary

This chapter, Chapter 5, presented and discussed the findings of the main phase of this study, including semi-structured interviews and a focus group discussion to validate the findings of these semi-structured interviews. Hence, the author provided the reader with a complete picture of this research project's findings that incorporated the translated narratives of the interview partners as an informative record. In this way, the author could present the development of the introduced concepts based on the informants' original expressions and discuss the theoretical backgrounds leading to the identified aggregated dimensions of Supply Chain Risk Management Process, Supply Chain Risk Management Culture and Supply Chain Risk Management Cooperation. The focus group discussion validating these findings helped to address the research question and its sub-research questions, which shows the need for a cooperative Supply Chain Risk Management approach.

The next chapter, Chapter 6, introduces the emergent theory of this research project by presenting a cooperative Supply Chain Risk Management approach for standardising or improving Supply Chain Risk Management processes and developing business opportunities. Based on this chapter's discussion of the findings, an interorganisational view of Supply Chain Risk Management is presented that once established in the organisational environment of the Hidden Champions studied, supports the transition of the existing firm-centric perspective to an interorganisational approach spanning the organisational boundaries of the Hidden Champions to mediate the management and mitigation the disruptive events in the upstream supply chain and ultimately increase supply chain resilience.

Supply Chain Risk Management—A Cooperative Approach

6

The previous chapter presented the findings from this research project based on qualitative data gained from semi-structured interviews with senior managers responsible for Supply Chain Risk Management, located mainly in purchasing or supply chain management of Hidden Champions in the DACH region. The identified aggregate dimensions from the data analysis of this research: Supply Chain Risk Management process, Supply Chain Risk Management culture and Supply Chain Risk Management cooperation were discussed in the light of existing research, validated through a focus group discussion, addressing the research question and sub-research questions and demonstrating the need for a cooperative Supply Chain Risk Management approach.

This chapter, Chap. 6, presents the emerging theory from this research project which serves the need for cooperative Supply Chain Risk Management by introducing an interorganisational view of Supply Chain Risk Management. Once established in the organisational environment of the Hidden Champions studied, the concept of cooperative Supply Chain Risk Management supports the transition of the existing firm-centric perspective to an interorganisational view of Supply Chain Risk Management spanning the organisational boundaries of the Hidden Champions to mediate the management and mitigation of disruptive events in the upstream supply chain and thus increase supply chain resilience. To standardise and improve the previously informal Supply Chain Risk Management processes of the Hidden Champions and that of their partners in the upstream supply chain, this chapter introduces a framework of a two-phase model towards a cooperative Supply Chain Risk Management process grounded in process improvement that is discussed based on the four-step plan-do-check-act cycle.

© The Author(s), under exclusive license to Springer Fachmedien Wiesbaden GmbH, part of Springer Nature 2026

M. Mayer, *Spanning Organisational and Geographical Boundaries*,
https://doi.org/10.1007/978-3-658-50453-3_6

6.1 Interorganisational View of Supply Chain Risk Management

The field research of this study has revealed a mostly firm-centric view of Supply Chain Risk Management that does not span the organisational boundaries of the Hidden Champions, given their mainly informal processes. This mostly firm-centric view implied that these Supply Chain Risk Management processes were subject to substantial isolation mechanisms (Peteraf 1993) and were only applied by the interviewed Hidden Champions themselves. These Supply Chain Risk Management processes were thus considered inimitable or isolated resources that were subject only to the Hidden Champions and not transferred to or shared with their supply chain partners (Barney 1991). However, the interorganisational view of Supply Chain Risk Management aims to combine the initially identified themes of training and development of the Hidden Champions' employees and Supply Chain Risk Management cooperation with being imitable and transferable practices to span organisational boundaries (Carter et al. 2017). In following Carter et al. (2017) further, extending the Supply Chain Risk Management practices to a dyadic or network-oriented buyer-supplier relationship complementing the Supply Chain Risk Management perspectives of the Hidden Champions with those of their supply chain partners would mediate the management and mitigation of disruptive events through their early identification. Hence, this interorganisational view would lead to more robust Supply Chain Risk Management and lower recovery cost (Sheffi and Rice 2005), or according to Ivanov (2020), would secure the viability of the Hidden Champions supply chain despite disruptive events.

Due to the increasing internationalisation of the Hidden Champions discussed by Audretsch et al. (2018) and the corresponding increase in the number of supply chain partners, the resulting increased complexity of the supply chain in connection with the informal Supply Chain Risk Management processes identified at the Hidden Champions assumes a high vulnerability of the upstream supply chain (Steinle and Schiele 2008). This increased complexity requires an interorganisational view that also incorporates the Hidden Champions' supply chain partners into the Supply Chain Risk Management processes by spanning the organisational boundaries.

An unpronounced Supply Chain Risk Management culture identified in the Hidden Champions of this study implies insufficient awareness of the upstream supply chain risks among their employees in the various functional areas besides those in purchasing and supply chain management. Further, there is a lack of understanding of the processes used to identify and avoid risks in the company. Appropriate training of the employees and defined talent management to address this lack of understanding does not take place or only to an insufficient extent. The isolated Supply Chain Risk Management only refers to the Hidden Champions' organisation, and the resulting idiosyncratic Supply Chain Risk Management processes among the supply chain members lead to a discrepancy in their approaches to Supply Chain Risk Management. However, interorganisational training that could harmonise the Supply Chain Risk Management approaches and processes among the supply chain members is not established. Reactive and thus ineffective communication with the supply chain partners in the event of supply chain disruption also hinders the early identification of disruptive events in the upstream supply chain, which leads to more difficult management and mitigation of such disruptive events, ultimately threatening the undisrupted flow of the Hidden Champions' supply chain. A company-wide Supply Chain Risk Management culture, supporting the same values and standards among the employees in all functional departments of Hidden Champions, leads to a high level of awareness regarding the risks in the upstream supply chain and the relevant processes that mediate the management and mitigation of disruptive events in the upstream supply chain. Through appropriate training and development, the employees of the Hidden Champions are adequately prepared for the ever-changing requirements in the more complex supply chains.

The theoretical framework emergent from this research illustrates the transition from a firm-centric Supply Chain Risk Management view identified in the field study of this research to an interorganisational view of Supply Chain Risk Management. Supply Chain Risk Management cooperation (1) can be established by implementing the process improvements presented in sect. 6.3. This cooperative Supply Chain Risk Management approach promotes the interorganisational view of Supply Chain Risk Management that enables the Hidden Champions and similar companies to adopt this perspective to dismiss the shortcomings of the identified firm-centric view of Supply Chain Risk Management and thus mediate the management and mitigation of disruptive events. The Hidden Champions can standardise their previously informal Supply Chain Risk Management processes and harmonise (2) these processes in cooperation with the processes of their supply chain partners and continuously improve them by applying the interorganisational view of Supply Chain Risk Management. In this way, the Hidden Champions' upstream supply chain members can cooperatively adapt to changes in their market environment and design their processes to be correspondingly efficient (Fig. 6.1).

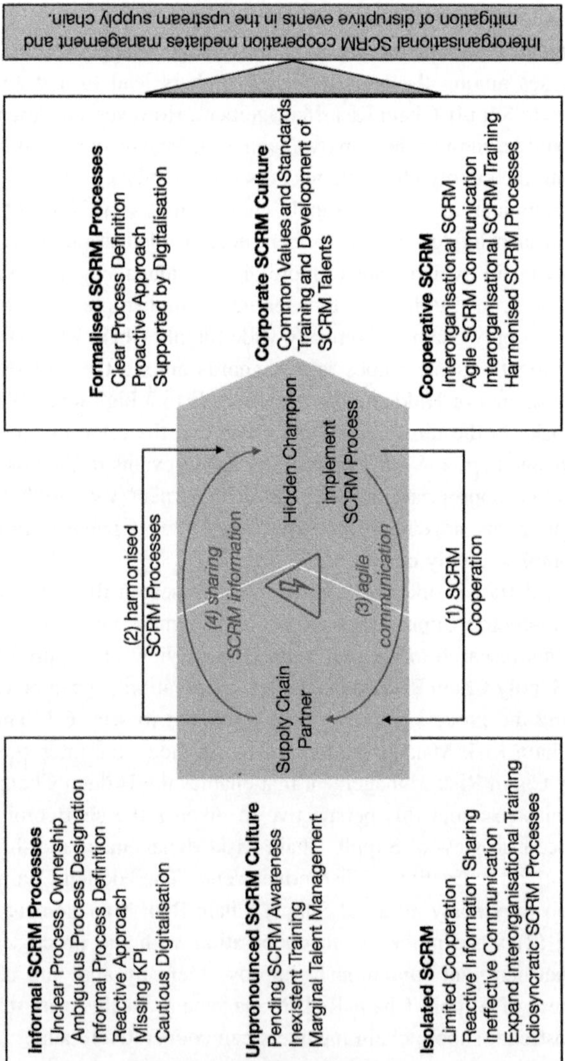

Fig. 6.1 Emergent Theoretical Framework of this Research

By adopting this proactive approach, Supply Chain Risk Management cooperation, supported by digital solutions where possible, Hidden Champions and similar companies can cooperate with their supply chain partners using agile communication (3) to identify disruptive events in the upstream supply chain at an early stage and by sharing Supply Chain Risk Management (4) information to effectively manage and mitigate them, thereby increasing the resilience and ultimately the performance of the upstream supply chain.

The following section presents an in-depth description of how Hidden Champions and similar companies can standardise and improve their Supply Chain Risk Management processes to achieve an interorganisational Supply Chain Risk Management view in cooperation with their supply chain partners.

6.2 Adopting a Cooperative Supply Chain Risk Management Approach

Due to the informal Supply Chain Risk Management processes at the Hidden Champions and the lack of knowledge about the Supply Chain Risk Management processes of their supply chain partners, this thesis presents a concept for standardising and improving those processes by adapting a cooperative Supply Chain Risk Management approach based on a continuous improvement process. Therefore, the author applies a process-oriented problem-solving approach, the iterative four-step plan-do-check-act (PDCA) cycle or Deming wheel (Saier 2017), a continuous improvement process, its origins and developments introduce Moen and Norman (2009) or Saier (2017) introduced in detail. This process centres on disseminating within the entire organisation of Hidden Champions, which often use process optimisation to maintain their leading position over competitors through innovative, easy-to-use processes as discussed by Moosa and Sajid (2010) or Schmieder (2018). The introduced plan-do-check-act cycle supports the optimisation of the Supply Chain Risk Management processes. It is suitable for implementation in Hidden Champions based on its simplistic design as requested by Zsidisin and Ritchie (2009), compared to other more complex and only project-oriented methods for process improvement that require intense training and qualification of Hidden Champions employees (Eckes 2001; Romdhane et al. 2017). This method allows implementing the cooperative Supply Chain Risk Management approach at a small number of supply chain partners and scaling up to a more significant number of supply chain partners in the subsequent iterations.

The following section explains the Supply Chain Risk Management process standardisation introducing the process design of the generic plan-do-check-act cycle based on which a conceptual framework of a two-phase iterative model towards a cooperative Supply Chain Risk Management process is developed.

6.3 Process Standardisation and Improvement

The underlying idea of standardising or improving the previously informal Supply Chain Risk Management processes of Hidden Champions is based on the process-oriented problem-solving approach, the plan-do-check-act cycle (Saier 2017). In the four steps depicted below, the cycle moves through these four steps to determine which specific aspects should be standardised or improved. The following section explains these four steps in detail. This plan-do-check-act cycle serves as the foundation for the two-phase model towards a cooperative Supply Chain Risk Management, see Fig. 6.2.

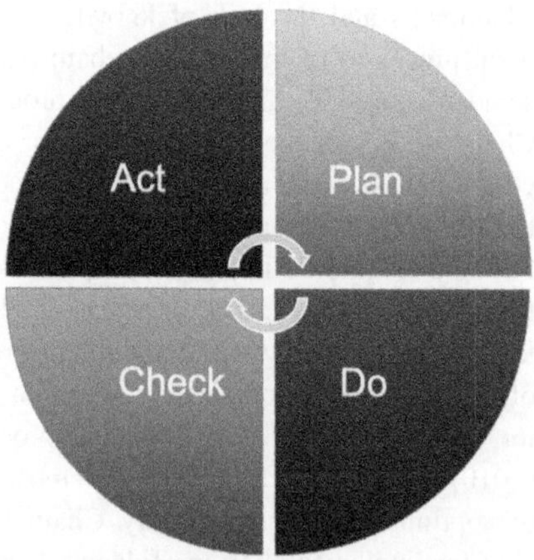

Fig. 6.2 Plan-Do-Check-Act Cycle. (Adapted from Sokovic et al. 2010)

The approach is based on a continuous improvement process that should be part of the organisational culture (Sokovic et al. 2010), its contents are explained following:

(1) Based on a detailed analysis of the current process, the Plan-step identifies the improvement potential of the process in use. The targets to be achieved in this process cycle are defined based on the earlier analysis.
(2) In the Do-step, the previously defined targets are implemented based on specified criteria.
(3) Within the Check-step, the earlier implemented targets are checked for effectiveness; if necessary, the corresponding activities are revised.
(4) The Act-step defines the newly implemented process as standard process and defines targets for the next process cycle, which improves this current standard process.

The individual steps to standardise and improve the Supply Chain Risk Management processes are introduced in the next section.

6.3.1 Two-Phase Model Process Model

The author recommends a two-phase process based on the Hidden Champions' Supply Chain Risk Management practices discussed during the semi-structured interviews to establish the proposed process of cooperative Supply Chain Risk Management in Hidden Champions and their supply chain partners. Also, the findings from the data analysis joining the interview partners' terms and expressions with theoretical insights lead to the development of this two-phase process model. In the first iteration, the Hidden Champions' previously informal Supply Chain Risk Management processes are standardised to eliminate the previously identified shortcomings and transform the corresponding Supply Chain Risk Management process into a stable state. The second iteration of the plan-do-check-act cycle aims to improve the above standardised Supply Chain Risk Management process of the Hidden Champions in cooperation with their supply chain partners to achieve greater transparency regarding the latter's Supply Chain Risk Management process and identify opportunities for mutual process improvement. Building on the plan-do-check-act cycle depicted above, a framework of a two-phase process model toward a cooperative Supply Chain Risk Management process is presented in the following sections.

6.3.2 Process Standardisation Cycle

This section focuses on the first iteration of the two-phase model to standardise the Hidden Champions' previously informal Supply Chain Risk Management process. The intraorganisational exercise of the Supply Chain Risk Management process through the employees involved and the corresponding identification of critical supply chain partners and their top-level decision-makers enable the cooperative Supply Chain Risk Management process to be standardised.

The *Plan-step to process standardisation* requests a detailed analysis of the current situation of the Hidden Champions' Supply Chain Risk Management processes. It aims at involving the Hidden Champions' employees to elaborate an improvement plan, including goals to be achieved (Al Smadi 2009). The objectives of the Plan-step are the development and implementation of cooperative Supply Chain Risk Management in Hidden Champions, the revision of the corresponding definitions and processes in line with the definition of cross-functional teams responding immediately in the event of a supply chain disruption as introduced by Travel (Travel 171) and discussed earlier by Norrman and Jansson (2004) or Helmold (2014), see Fig. 6.3.

The involvement of employees in this process improvement is intended to increase acceptance of the Supply Chain Risk Management process and thus risk awareness, thereby establishing a Supply Chain Risk Management culture in the companies leading to increased resilience (Ates and Bititci 2011). Training concepts shall be implemented for employees to participate actively and thus contribute to process improvement through their experience. Thereby advancing the comprehensive identification of critical supply chain partners further contribute to establishing the required culture; their needs are only recognised by some interview participants (Lift 23; Cooper 14). Employing a top-level communication channel in case of disruptive events in the upstream supply chain to promote agile communication, as promoted by Wieland and Wallenburg (2013) or required by Bals and Turkulainen (2019), is supported by the close and long-term partnership of the Hidden Champions with their supply chain partners as confirmed by Mill (Mill 21).

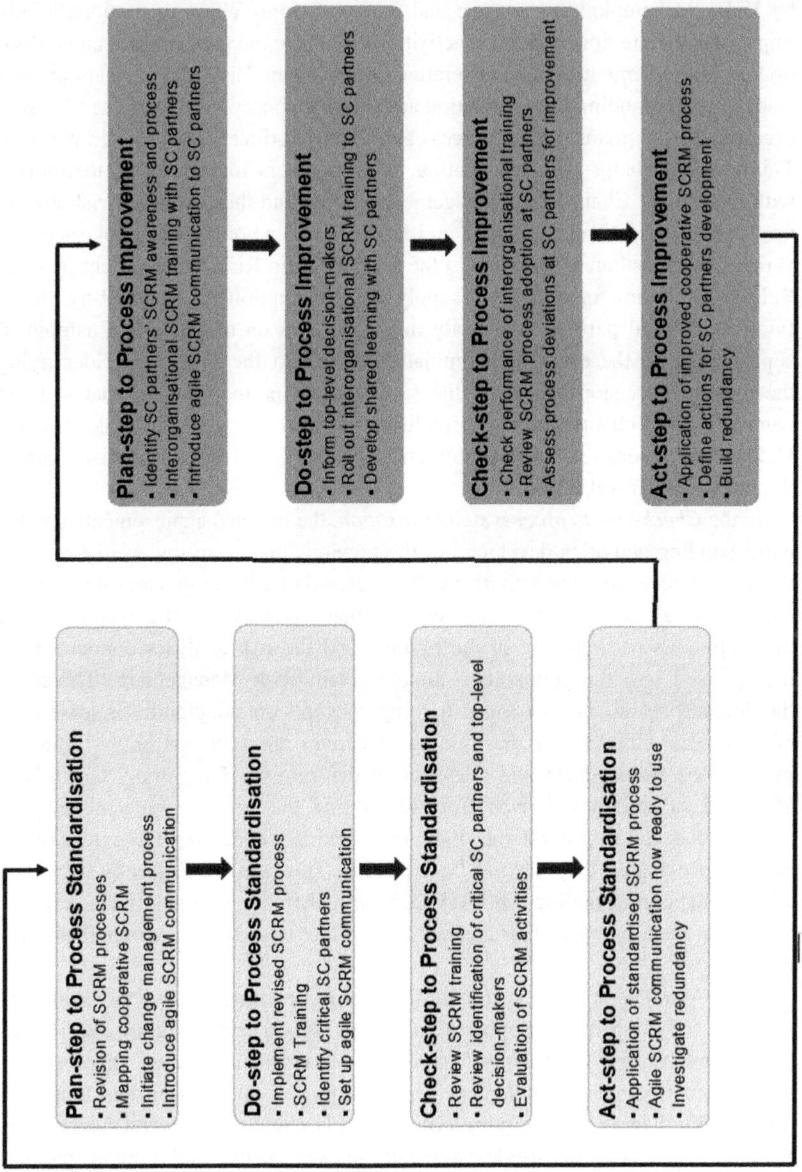

Fig. 6.3 Process Standardisation Cycle

In the *Do-step to process standardisation*, the cross-functional teams defined by Hidden Champions to manage and mitigate supply chain disruptions jointly implement the previously defined activities to achieve the goals of the standardisation cycle. Implementing the cooperative Supply Chain Risk Management process leads to corresponding documentation and training concepts for later use to jointly exercise the risk management process and identify critical supply chain partners. This approach helps the participating team members to familiarise themselves with the Supply Chain Risk Management process and thus promotes risk awareness and team building (Tuckman and Jensen 1977) among the employees of the various functional areas involved in the Supply Chain Risk Management process. Relevant activities for mitigating supply chain disruptions are defined by identifying the critical partners and analysing the effects on the Hidden Champions' supply chain in the event of a simulated failure. At the same time, identifying the top-level decision-makers at the supply chain partners means that a direct communication channel can be established between the decision-makers at the Hidden Champions and their supply chain partners, as Travel recognised during the interview (Travel 329).

In the *Check-step to process standardisation*, the targeted achievements and the corresponding activities developed in the previous phase are reviewed for effectiveness; if necessary, the activities will be revised to achieve greater effectiveness (Sokovic et al. 2010). This review ensures that feedback and the experiences of the employees participating in the training and the risk analyses conducted are incorporated into the cooperative Supply Chain Risk Management. Therefore, the documentation and associated training concepts are adapted to increase their effectiveness. Checking if the training objectives are achieved helps further to increase the effectiveness and relevance of the training. The Supply Chain Risk Management activities derived from identifying the critical supply chain partners are checked to see whether they are sufficient concerning the extent of the damage or whether they need to be expanded. This step concludes by reviewing whether the corresponding top-level decision-makers at the supply chain partners have been identified and their contact details have already been comprehensively confirmed.

In the *Act-step to process standardisation*, the revised process represents the standard for cooperative Supply Chain Risk Management within the Hidden Champions. To use the Supply Chain Risk Management process appropriately, the Hidden Champions implement corresponding documentation and employee training. Furthermore, identifying critical supply chain partners and the communication channels to the top-level decision-makers at the supply chain partners support the implementation.

The following iteration further improves the cooperative Supply Chain Risk Management process and extends beyond the boundaries of the Hidden Champions to their supply chain partners. As a result of this improvement cycle, cooperation with the supply chain partners is established.

6.3.3 Process Improvement Cycle

This section focuses on the second iteration of the plan-do-check-act cycle to improve Hidden Champions' Supply Chain Risk Management process in cooperation with their supply chain partners to achieve greater transparency regarding the latter's Supply Chain Risk Management process and identify opportunities for mutual process improvement (Fig. 6.4). Priority within this process improvement cycle is given to those supply chain partners identified as critical in the analysis of the Supply Chain Risk Management processes and those having a significant effect on the Hidden Champions supply chain in the event of a simulated failure. In the *Plan-step to process improvement,* the cooperative Supply Chain Risk Management process, initially implemented at Hidden Champions, is extended beyond organisational boundaries to the supply chain partners.

Just as supported by Conche (Conche 10), this improvement cycle aims to involve the supply chain partners in the cooperative Supply Chain Risk Management process. In this way, Hidden Champions and their supply chain partners can support each other in managing and mitigating supply chain disruptions, either through cooperation on an operational level (Thomson and Arney 2015) or on a financial level (Tate et al. 2019). Both forms of cooperation offer excellent potential for improving Supply Chain Risk Management and thus increasing resilience, hence the performance of the upstream supply chain. This improvement cycle's objective is to identify the risk awareness of the employees and identify the Supply Chain Risk Management processes within the supply chain partners. Integrating the earlier defined training concepts into the cooperative Supply Chain Risk Management process also improves the risk awareness of the supply chain partners, state Kumar and Anbanandam (2020). At the same time, the cooperative Supply Chain Risk Management training of supply chain partners also helps to identify their existing Supply Chain Risk Management processes. Conche confirms that in this way, the Hidden Champions and their respective supply chain partners can learn from each other's experiences and use these experiences to optimise their Supply Chain Risk Management processes (Conche 86). Once the training concept supports identifying critical issues for the supply chain members, this offers the opportunity to provide the supply chain partners with development

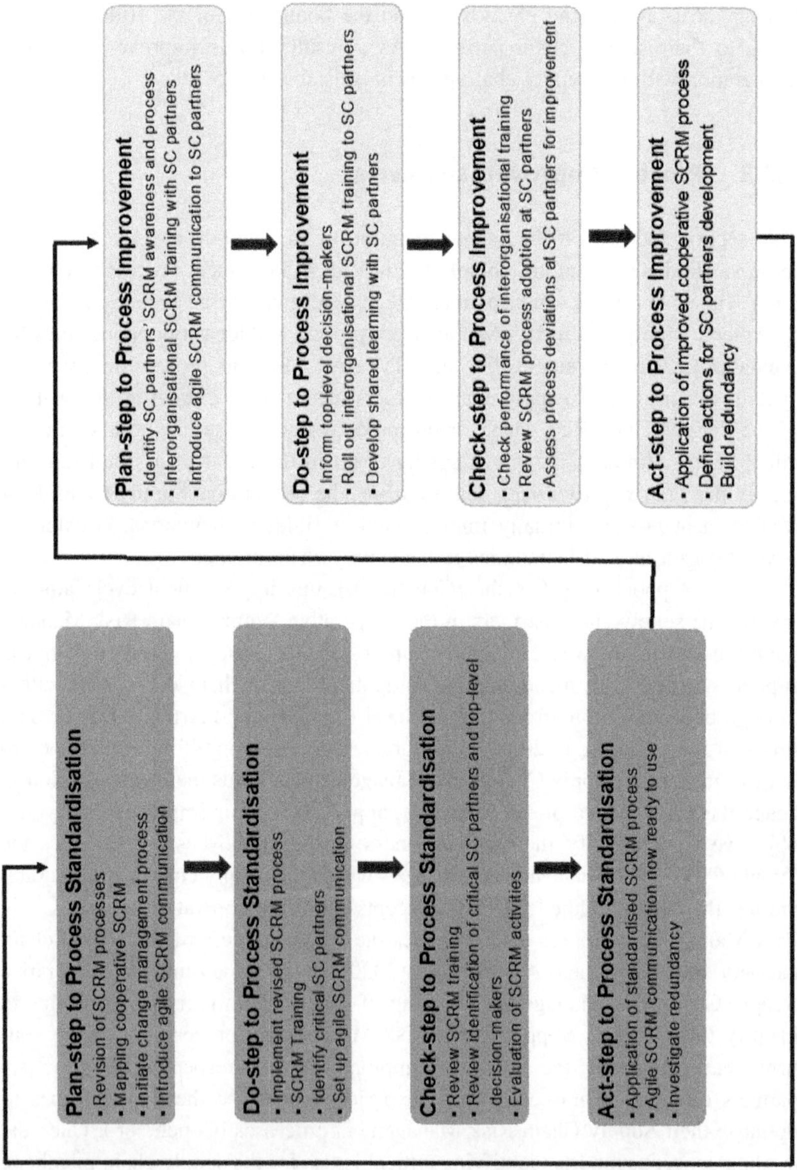

Fig. 6.4 Process Improvement Cycle

approaches or, if necessary, to increase redundancy in the supply base. To establish a direct communication channel to the identified top-level decision-makers of the supply chain partners, these should be informed of this approach and the possibilities of agile communication to promptly eliminate disruptive events in the upstream supply chain, which Open welcomed (Open 118).

In the *Do-step to process improvement*, the Hidden Champions extend the Supply Chain Risk Management training concept to their supply chain partners by informing their top-level decision-makers about the initiative to use the training concepts; however not always met with approval as put forward by Mill (FGD 11). Crises, such as the Covid-19 pandemic, have demonstrated the greater need for cooperative Supply Chain Risk Management training to create awareness of risks and potentially disruptive events in the upstream supply chain (Bessant et al. 2003). The awareness of disruptive events is an opportunity to motivate supply chain partners to engage in cooperative Supply Chain Risk Management training, which would not have been possible before such a crisis. This awareness leads to the top-level decision-makers of the supply chain partner being sensitised to the necessity of cooperative Supply Chain Risk Management, which motivates them to establish a direct communication channel. Once the cooperative Supply Chain Risk Management training is implemented at the supply chain partners, the process can be carried out in the companies and together with the supply chain partner's employees to learn from each other's experiences and use these experiences to mutually optimise the Supply Chain Risk Management processes which Experiment presented (Experiment 79).

In the *Check-step to process improvement*, the results developed with the supply chain partners in the previous step are analysed by the Hidden Champions. If the activities for achieving the aim to involve the supply chain partners in the cooperative Supply Chain Risk Management are sufficient, and all relevant supply chain partners participated in the training, no further improvement is necessary in the next step. Furthermore, it is reviewed whether risk awareness is established among the supply chain partners. While this risk awareness does not always seem to exist, as seen by Lift (FGD 99), Supply Chain Risk Management processes must be established and integrated into the adjunct business processes. If deviations from previously defined targets are identified, those need to be improved during the next phase with appropriate actions.

In the *Act-step to process improvement*, the cycle's last step, the corresponding improvement potential identified in earlier steps during the cooperative Supply Chain Risk Management training modules is now implemented and documented in the supply chain members' processes to manage and mitigate risks. The approach to further Supply Chain Risk Management development of the supply chain partners is supported from a theoretical perspective (Helmold (2014) and finds agreement with the interview participants. However, it is noted that the development of supply chain partners is very time-consuming (Experiment 198). Suppose the Hidden Champions identified deviations in their partners' Supply Chain Risk Management processes requiring further support. In that case, the Hidden Champions should design an extended action plan to support the supply chain partners in standardising or improving that process. If, after following this comprehensive action plan, there are still critical findings occurring at the supply chain partners, the resilience of the upstream supply chain can be increased through investments in redundancy, as proposed by Sheffi (2019). After completing the process revision at the Hidden Champions and the supply chain partners, the first iteration of the process improvement cycle ends. Once opportunities for improvement are identified in the further use of the cooperative Supply Chain Risk Management process, a new improvement cycle is initiated. The following section discusses the approach presented for standardising and improving the Supply Chain Risk Management process and differentiates it from existing frameworks requesting risk management.

6.4 Discussing the Emergent Framework

To provide a comparison of the emergent framework of this research with other existing frameworks for risk management, this section discusses a frequently used framework found in the ISO 9001:2015 standard (ISO 2015) applied by many businesses in the cross-border area of the DACH region and beyond. Based on this comparison, the following section discusses the identified differences.

6.4.1 Formalised Supply Chain Risk Management Processes

This research has identified the need for formalised Supply Chain Risk Management processes in which tasks and process ownership are clearly defined. The formalisation required leads to consistent terminology throughout the organisation; lacking formalisation and thus inconsistent terminology is seen as the primary inhibitor to the implementation of effective Supply Chain Risk Management (Diehl and Spinler 2013). Through clearly defined processes supported by digital solutions to take over administrative activities, a uniform understanding can be created to achieve a proactive approach to managing and mitigating disruptive events in Hidden Champions' Supply Chains. The above request for formalisation stands in contrast to the international standard ISO 9001:2015 (ISO 2015), which calls for measures to address risks and opportunities in the company in general (ISO 2015, p. 8); however, the international standard does not require formalised risk management processes, nor does it discuss a dedicated Supply Chain Risk Management process. Furthermore, the decision to document and formalise the Risk Management process lies in the hands of the organisation implementing the standard (ISO 2015, p. 53). However, the ambiguities identified in this thesis concerning process ownership, designation and definition of Supply Chain Risk Management processes showed that it is essential to create a clear understanding of Supply Chain Risk Management within the organisation.

6.4.2 Corporate Supply Chain Risk Management Culture

This research has further identified the need for a corporate Supply Chain Risk Management Culture based on common values and standards built on the formalised processes mentioned earlier. Enabling all members of an organisation to have shared values and leads to elevated levels of awareness regarding the risks in the upstream supply chain among all of the distinct organisational functions. These elevated levels of awareness are supported further by the continuous training and development of talents in the Supply Chain Risk Management function leading to a robust and stable organisation being able to master the challenges of constant change in the supply chain and retain talents and thus knowledge in the organisation. Within the framework of ISO 9001:2015 (ISO 2015), organisations are required to ensure that their employees are aware of the call for risk-based thinking. However, explicit supply chain risk awareness, as expressed in this thesis, to establish risk awareness resulting in a supply chain risk management culture within the entire organisation and across organisational boundaries to propagate in the organisations of the supply chain members is not required in the international standard. A further crucial issue for the Hidden Champions is their lack of sufficient skilled employees, especially in the Supply Chain Risk Management function. However, the international standard discusses to provide the relevant persons that master the implementation and the usage of the standard. The international standard requires organisations further to ensure suitable actions, such as training of employees, to provide knowledge and increase the competence of the employees as it is recognised to foster the effective usage of tools and processes wherever existing. However, as outlined in this research's emerging framework, clearly defined processes for knowledge transfer on a cross-functional level and managing talent within the organisation remain unaddressed in the international standard.

6.4.3 Cooperative Supply Chain Risk Management

Building on the organisation's formalised Supply Chain Risk Management processes and inherent culture, organisations should strive to apply these processes in such a way that the former idiosyncratic Supply Chain Risk Management processes are implemented, spanning organisational boundaries to the partners in the upstream supply chain. This interorganisational approach enables agile communication between decision-makers of the members in the upstream supply chain to identify disruptive events in the supply chain in good time and deal with them immediately. To increase supply chain members' awareness of risks and processes, this awareness should be raised through interorganisational training based on harmonised Supply Chain Risk Management processes.

Where the ISO 9001:2015 standard requests a customer focus to address risk threatening the ability to deliver to the customer, cooperation to manage and mitigate risks in the supply chain is not in the scope of the international standard. While agile communication facilitates the information flows in the upstream supply chain, only communication with the customer (ISO 2015, p. 32) is discussed in the context of communication. This communication is undoubtedly of immense importance in the context of cooperative Supply Chain Risk Management, as it helps organisations quickly identify risks and manage them in a targeted manner. However, direct contact between the decision-makers of the upstream supply chain is not discussed in the international standard, which can lead to delays in the process, where agility would be of high importance

As mentioned in the previous section, the international standard addresses the need to train employees to increase their knowledge and competence within the organisation. An interorganisational training to share the requirements for Supply Chain Risk Management among the supply chain members is not foreseen. Therefore the international standard describes (ISO 2015, p. 8) its common use by internal and external partners. Furthermore, it requires the control of externally provided processes (ISO 2015, p. 38) to meet the organisation's requirements. A harmonisation of Supply Chain risk management processes, including the cooperative adaptation of processes to create a common understanding enabled through mutual learning, which ultimately improves collaboration as proposed in this thesis, extends the requirements of the international standard.

6.5 Chapter Summary

By presenting the theory emerging from this research project in this chapter, the author illustrated the relevance of cooperative Supply Chain Risk Management. The introduced concept of cooperative Supply Chain Risk Management supports the transition of the existing firm-centric perspectives to an interorganisational view of Supply Chain Risk Management, spanning the organisational boundaries of the Hidden Champions to their supply chain partners. Therefore, a two-phase model towards a cooperative Supply Chain Risk Management process grounded in process improvement following the plan-do-check-act-cycle was established to standardise and improve the previously informal Supply Chain Risk Management processes of the Hidden Champions and their partners in the upstream supply chain. A discussion with the existing framework of the ISO 9001 standard has identified the need for such as framework while it extends the requirements of currently existing frameworks. Based on its simplistic design, the cooperative Supply Chain Risk Management process is suitable for implementation in Hidden Champions and similar companies and contributes to their benefit. Other more complex and only project-oriented methods for process improvement require intense training and qualification of Hidden Champions employees. This simplistic design allows the Hidden Champions to start implementation with a small number of supply chain partners and scale up to a more significant number of supply chain partners in subsequent improvement cycles.

The concluding chapter, Chap. 7, will provide this thesis's summary. By reflecting on the research process and based on the emergent theory of this research, the author presents the contribution of this project to theory and practice. The limitations of this research produce recommendations that future researchers could investigate in. The last section of this concluding chapter provides an overview of current and future developments in risk management in the context of increasing challenges in the global supply chain.

7 Conclusion and Recommendations for Future Work

A final discussion of the cooperative approach to Supply Chain Risk Management will specifically address this research's question and sub-research questions, followed by a personal reflection of the research methodology used, which initiates this concluding chapter. By presenting the contribution to knowledge from a theoretical and a practical perspective, the author demonstrates the originality and significance of this work for academia and practitioners. This is followed by a discussion of the limitations of this research project, helping the reader comprehend the recommendations for further research, in that this thesis will be addressing the methodological and content-related areas for future research. The last section touches on the recommendations of this research in the light of current and future developments in Supply Chain Risk Management.

7.1 Findings in Relation to the Research Questions

The initial literature review introduced in Chap. 2, prefacing this study, presented a research gap that is significant for the current research on Supply Chain Risk Management in Hidden Champions in the mechanical engineering industry in the cross-border area of the DACH region. This research gap shows that Supply Chain Risk Management within Hidden Champions has not been a topic of academic research before. The identified gaps in academic research are the lack of understanding of the cooperation between the Hidden Champions and their supply chain partners to manage and mitigate disruptive events in the upstream supply chain, requiring intensive communication and instant information sharing. Furthermore, it was unclear from the literature review how Hidden Champions are empowering their supply chain partners and whether the existence of a Supply Chain Risk Management process at these partners would mediate or positively

© The Author(s), under exclusive license to Springer Fachmedien Wiesbaden GmbH, part of Springer Nature 2026

M. Mayer, *Spanning Organisational and Geographical Boundaries*,
https://doi.org/10.1007/978-3-658-50453-3_7

influence the management and mitigation of disruptive events throughout the Hidden Champions' upstream supply chains.

These research gaps have led to the definition of the following research question for this study:

How does interorganisational cooperation between Hidden Champions and their suppliers mediate the management and mitigation of disruptions occurring in the upstream supply chain of Hidden Champions in the DACH region?

By addressing the following two constituent sub-research questions, this research question can be answered in full:

i. How are the Hidden Champions cooperating with their supply chain partners to manage risks in the Hidden Champions' upstream supply chain?
ii. How are the Hidden Champions employees and the supply chain partners trained and developed to understand and apply the Supply Chain Risk Management processes available?

Based on the findings in Chap. 5, that discussed the findings of this research the author can confirm that interorganisational cooperation mediates the management and mitigation of disruptive events in the upstream supply chain of Hidden Champions in the DACH region. Crucial for managing and mitigating disruptive events in the upstream supply chain of the Hidden Champions is the close and long-standing cooperation of the Hidden Champions with their supply chain partners leading to mutual support in case of disruptive events in the upstream supply chain. This kind of relationship-inherent Supply Chain Risk Management is challenging to replace by other mechanisms as it aims at the well-being of all members of the upstream supply chain. In this way, Open described a situation with a supply chain partner that goes back decades, where they supported the partner in distress, which still positively affects the close relationship between the two companies today:

> *"One of our biggest suppliers was in trouble in the 80s and we helped them financially and helped them to recover and today it is one of our best suppliers and completely independent, expanding and happy and satisfied. Because of this cooperation, I have a great working relationship on this side."* (Open 112).

Given the positive impact on supply chain resilience, this close cooperation in managing and mitigating risks, however, does not follow any recognisable systematic approach, nor are the processes to be used or the objectives to be achieved harmonised or coordinated in any way in advance. The interview partners supported employee training for them to better understand the Supply Chain Risk Management requirements in advance, as confirmed by Conche:

> *"No, for me, it's actually about qualifying them* [the employees], *to train them."* (Conche 90).

However, training the employees and supply chain partners to increase Supply Chain Risk Management cooperation requires stronger emphasis. Therefore, Lift requests training of the supply chain partners,

> *"... where we explain strategies and the expectations of how we can work better together."* (Lift 71).

The design of the Supply Chain Risk Management process at the Hidden Champions considered in this study is still in its infancy. Risk analyses are created and documented using spreadsheet programmes, as reported from Mill, among others:

> *"In the end, we built risk management in such a way that we simply made an Excel table."* (Mill 24).

Hence system-guided Supply Chain Risk Management depends on the prerequisites of the enterprise resource planning system used or is currently being developed. The use of digital Supply Chain Risk Management solutions was also discussed, although it was unclear which added value these solutions could offer, also against the backdrop of the implementation costs, which were described as extremely high.

Activities that supported dealing with disruptive events that occurred in the upstream supply chain of many Hidden Champions were cross-functional task forces, as illustrated by Travel:

> *"For us, if something goes wrong or doesn't work, if we see a risk, also from the top management, then we say here we'll set up a task force with all the disciplines involved, with someone who has the hat on, that's what we've done again and again."* (Travel 170).

Therefore, a lively Supply Chain Risk Management culture, as demanded by the interview partners, only exists in very few cases. See Experiment's request for example:

> *"The topic of risk management must be much more strongly anchored and perceived in the players' awareness."* (Experiment 26).

Employees in purchasing and supply chain management need training on Supply Chain Risk Management, as well as training between the individual participants in the task force is required to establish both team building and risk awareness, which could ultimately lead to a lively Supply Chain Risk Management culture in the company.

To further promote cooperative Supply Chain Risk Management to manage and mitigate disruptive events in and thus increase the resilience of the upstream supply chain, Conche proposes the cooperative Supply Chain Risk Management in firms similar to the Hidden Champions studied:

> *"But what I imagine, in a team of companies that have the same mindset, I could very well imagine that you could work together."* (Conche 21).

In summarising, the author can state that the findings discussed in relation to the research questions addressed those affirmatively. The Hidden Champions participating in this study support the presented cooperative approach to Supply Chain Risk Management. However, implementation in the studied Hidden Champions is still pending. To support this implementation, the presented process to standardise and improve the current Supply Chain Risk Management approaches can be used.

7.2 Personal Reflection

This research project started in April 2015. At that time, Supply Chain Risk Management research had already found its way into academia for almost two decades. Among practitioners, Supply Chain Risk Management research found only minor interest. Then, the supply chain events driven by the Covid-19 pandemic partially brought international and regional supply chains considered highly resilient to a standstill. Hence, Supply Chain Risk Management to manage and mitigate disruptive events in the supply chain has become a prominent factor.

This project was a significant gain for the author. On the one hand, by gaining on a personal level being able to accomplish this research project despite the threefold challenge between doctorate, teaching as a senior lecturer in purchasing and supply chain management and of course, the family obligations of being a son, brother, husband, and family dad of a family with four small children. On the other hand, by gaining professionally through applying an interpretive research approach. The interpretive approach allowed the author to learn about qualitative research methods and apply them successfully, enabling a deep immersion in the subject area of Supply Chain Risk Management at Hidden Champions. This deep immersion in the subject area was beneficial for the author's teaching at Universities in business administration or engineering and management in the DACH region, where the internationalisation of the supply chain of often small and medium-sized family businesses is of immense importance.

Furthermore, the interpretive research approach chosen was crucial to the success of this project, as the deep insight into the Supply Chain Risk Management approaches at the Hidden Champions would not have been possible otherwise. By using semi-structured interviews, the author gained rich information through face-to-face conversations with the interview partners, which implied a significant investment of time and resources because of travelling to the interview partners' locations. Through more effective use of electronic communication solutions, this time expenditure could have been significantly reduced. However, electronic communication would have borne the risk that such a deep immersion into the topic would not have been possible. When the author had to use electronic communication, the interaction with the interview partner was limited to verbal communication, while non-verbal communication could not be obtained, which narrowed the richness of information obtained from the interview. The coding and analysis of the qualitative data using the Gioia Method (Gioia et al. 2013) were also time-consuming; nevertheless, the excellent preparation and the systematic analysis of the information from the semi-structured interviews supported the subsequent focus group discussion to validate the analysis' results providing qualitative rigour, delivering plausible and defensible conclusions.

7.3 Contribution to Knowledge

The originality of this work builds on the critical evaluation of existing research in the topic areas of Hidden Champions and Supply Chain Risk Management and provides an overview of the success strategies of Hidden Champions. The topic area of Supply Chain Risk Management was identified as a gap in the knowledge of Hidden Champions and was presented accordingly. To investigate the use of Supply Chain Risk Management in Hidden Champions, an interpretive research approach was used to support this study methodologically. This research approach aimed to identify new theoretical insights and thus explained the phenomenon of Supply Chain Risk Management in the Hidden Champions participating. The chosen interpretive research approach extends the scientific knowledge by using a philosophical approach which is not common in Supply Chain Risk Management research. The explanatory attitude of this research project demonstrates the importance of a cooperative Supply Chain Risk Management, previously not identified with the majority of studies in this discipline adopting a quantitative approach.

7.3.1 Contribution to Theory

The theoretical contribution of this explanatory study of Supply Chain Risk Management in Hidden Champions in the mechanical engineering industry in the cross-border area of the DACH region is based on the initial literature review presented in Chap. 2. It contributes to the literature by providing an initial review of the extant literature on Hidden Champions and Supply Chain Risk Management; this research highlights the economic and social importance of Hidden Champions for the cross-border area of the DACH region; it further identifies the challenges for supply chains arising from the globalisation of Hidden Champions. Despite an increasing scarcity of skilled employees, the Hidden Champions show a high resilience to disruptive events in the supply chain, but at the same time, the initial literature review identified ambiguous definitions of SCRM, which impede the implementation of clearly defined Supply Chain Risk Management processes as such and in cooperation with the supply chain partners. Therefore the author provided a definition for cooperative Supply Chain Risk Management leading to close cooperation and a vital communication culture with the supply chain partners, which are of immense importance in recognising supply chain disruptions in good time and dealing with them immediately. With no definition for cooperative Supply Chain Risk Management existing previously, the author thereby contributes to the theoretical knowledge.

Applying the supply chain practice view, as a theoretical lens, that uses the dyad or network as the organisational level of analysis further contributes to Supply Chain Risk Management literature. Thus, unlike the resource-based view or the socioemotional wealth theory, theoretical approaches that prevail in the identified literature, the supply chain practice view spans across organisational boundaries and includes the Hidden Champions' partners in the upstream supply chain in the analysis. Applying the supply chain practice view leads to increased performance across the entire supply chain rather than only creating sustained competitive advantage for the company leading the supply chain. This interorganisational level of analysis is applied to Supply Chain Risk Management in the context of this study. Thus, the analysis of the collected data shows that interorganisational cooperation between Hidden Champions and their supply chain partners mediates the management and mitigation of supply chain risks. The supply chain practice view, unlike the resource-based view, uses interorganisational practices and resources that are imitable and transferable; it enables all partners in the supply chain to apply Supply Chain Risk Management practices and thus jointly identify risks in good time and thereby improve the performance of Supply Chain Risk Management and thus the supply chain's overall performance (Pagell et al. 2022).

Highlighted by the supply chain practice view as an interorganisational practice is knowledge sharing or information sharing that was identified in the discussion of the findings. Where supply chain partners proactively shared information between the upstream supply chain members, thus increasing resilience by early identification of disruptive events, those were awarded new business preferential over others (Glide 117). Supply chain partners that did not show such cooperative behaviour were phased out at the earliest possible opportunity (Mill 45). A further transferable practice identified during the discussion of the findings, was interorganisational cooperation, where the members of the Hidden Champions' supply chains supported each other, either through the transfer of resources such as staff secondment (Travel 321) or through financial support (Irrigate 180).

The specific Supply Chain Risk Management training and development of Hidden Champions' and supply chain partners' employees identified during the initial literature review that will support cooperation and mutual understanding of Supply Chain Risk Management processes, however, needs further development to enhance Supply Chain Risk Management performance. Through the presented model towards an interorganisational Supply Chain Risk Management crossing organisational boundaries and thus creating opportunities for a cooperative approach to Supply Chain Risk Management, this work contributes to theory by demonstrating how the application of the supply chain practice view on Supply Chain Risk Management can generate increased performance in the supply chain.

7.3.2 Contribution to Practice

In practice, the topic of Supply Chain Risk Management is vital to Hidden Champions for maintaining operational performance expressed through customer satisfaction through high delivery reliability and holding strong market positions against competitors by establishing cooperative Supply Chain Risk Management. This cooperative approach is already well established among many of the Hidden Champions participating in this study, even though not formalised. The Hidden Champions have long-standing relationships with their supply chain partners, which creates close ties and intense loyalty (Mill 21).

While the semi-structured interviews and the systematic qualitative analysis did not provide generalisable results from a statistical point of view, the presented approach of a cooperative Supply Chain Risk Management can be established in companies of varied sizes and from different sectors to standardise the Supply Chain Risk Management processes and to continuously improve them in cooperation with their supply chain partners. Developing a concept towards a cooperative Supply Chain Risk Management becomes evident through the resulting standardisation and subsequent improvement of Supply Chain Risk Management processes in the Hidden Champions studied and at their supply chain partners. Based on its simplistic design, the opportunity to include a significant number of supply chain partners and the ease of use contributes to the benefit of the Hidden Champions. Their Supply Chain Risk Management processes were not formalised before and thus helped to increase resilience in the Hidden Champions' upstream supply chain. This cooperative approach to Supply Chain Risk Management shall be seen as an approach within the entirety of the considered enterprise rather than focusing solely on the purchasing and supply chain management organisation.

Higher risk awareness is generated through improved Supply Chain Risk Management knowledge of the supply chain members, which helps to identify a more extensive number of supply chain risks and define better mitigating strategies in case of supply chain disruption. This way, the affected company can initiate a faster response in cooperation with its supply chain partners to deal with the disruptions immediately. Therefore, cooperation with supply chain partners is vital to the Hidden Champions, creating close relationships between the supply chain members. Clearly defined contact persons and communication channels help initiate faster and more sustainable responses to alert the upstream supply chain earlier and thus increase resilience. In times of material scarcity or similar disruptive events in the upstream supply chain, quick reactions improve the performance of Supply Chain Risk Management and create advantages over competitors.

Through cooperative Supply Chain Risk Management, employee training in the involved departments is encouraged, which improves awareness leading to a proactive Supply Chain Risk Management culture, prompting faster reactions to disruptive supply chain events. Furthermore, Supply Chain Risk Management training supports the development and retaining of talents for the companies by increasing the purchasing and supply chain management organisation's visibility in applying a cooperative Supply Chain Risk Management approach. This joint elimination of disruptive events across organisational boundaries increases the resilience of the Hidden Champions' upstream supply chain, hence the operational performance towards the customer. With this transformation towards cooperative Supply Chain Risk Management, supported by training and development, a change process is being initiated in many purchasing and supply chain management organisations, which supports businesses to deal with the developments in their globalised supply chains and thus find solutions for the upcoming challenges through innovative processes.

Furthermore, this thesis contributes to practice by supporting businesses to cope with the latest requirements of the German Supply Chain Act, approved by the German Federal Cabinet as *Act on Corporate Due Diligence Obligations in Supply Chains* (2021) or the future requirements of the related *European Union* (EU) *Supply Chain Directive.* The proposal for an EU Supply Chain Directive will require all businesses in the European economic area, including the businesses in the DACH region considered in this thesis, to establish risk management to identify social and environmental risks besides the economic risks already identified. This legal framework shall help to globally protect workers and the environment by creating a framework for due diligence along the supply chain. The German Supply Chain Act and the future EU Supply Chain Directive are particularly important for the participating businesses in this study and comparable businesses. The German Supply Chain Act considers the German medium-sized businesses, requiring businesses with generally more than 1,000 employees to review their global value chains and implement the legal obligations by 1 January 2024. According to the European Commission (EC), the future EU Supply Chain Directive extends this scope and includes businesses with more than 500 employees or more than 250 employees when operating in a high-impact sector such as the minerals, clothing or agro-food sector and is expected to be ratified earliest in 2024 (EC 2022).

7.4 Limitations and Recommendations for Future Research

This section describes the limitations of this research project, conducted as an interpretive study identifying Supply Chain Risk Management approaches of Hidden Champions in the mechanical engineering industry in the cross-border area of the DACH region. The limitations indicated give recommendations for further studies in Supply Chain Risk Management in Hidden Champions to encourage future researchers to investigate the topic further.

7.4.1 Limitations of this Research

The geographic focus of this study lies in the concentration of Hidden Champions in the DACH region, where there is a substantial occurrence of Hidden Champions with similar cultural backgrounds. Thus, this study is limited to specific companies in a specific industry sector of a geographical region. This study is further limited to interorganisational cooperation to manage and mitigate risks in the upstream supply chain. Hence, this thesis has only marginally considered the topic of Supply Chain Risk Management Talent Management with the adjacent topic area of Supply Chain Risk Management Knowledge Management, as this is a topic area that could not have been conclusively dealt with within the scope of this thesis.

This study would not have been successful without the advantage of interpretive research to understand and explain interorganisational relationships. For example, the proximity to the interview partners and the immersion in the subject area would not have been possible through a quantitative approach such as using a questionnaire. On the other hand, and this is the limiting element of the applied interpretive approach, this study does not aim to provide statistical generalisable results but verify how the capabilities of awareness, processes and cooperation with the supply chain partners support the management and mitigation of disruptive events in the upstream supply chain of Hidden Champions.

7.4.2 Recommendations for Future Research

This section describes areas for future researchers to overcome the limitations of this research project described in the previous section. The recommendations for future research are divided into a methodological part that describes further research avenues from a methodical point of view and a content-related part that introduces further research possibilities from a content point of view.

The methodical part of the recommendations for future research is based on this project's adopted interpretive research approach to investigate the application of Supply Chain Risk Management in Hidden Champions. This explanatory, theory- generating approach addressed the research question of this study. However, the applied approach did not provide statistically generalisable results to confirm this theory. By enlarging the sample size, future researchers can use quantitative methods to test the theory of cooperative Supply Chain Risk Management by examining the effects of cooperative Supply Chain Risk Management on the performance of the supply chain and thus on organisational performance. Furthermore, it can also be tested if cooperative Supply Chain Risk Management has a pro-competitive effect. Depending on the research approaches of future researchers, the corresponding methods would then be derived from these research approaches. For example, future researchers could adopt a mixed methods approach to research as introduced by Creswell (2014) or Saunders et al. (2009), using either in parallel or sequential qualitative and quantitative research methods. Based on this, future researchers could select a case study approach (Eisenhardt 1989, Yin 1984) in which the researcher identifies a company that supports the implementation of the presented cooperative Supply Chain Risk Management approach to standardise and improve its Supply Chain Risk Management processes and those of their supply chain partners. By using qualitative methods such as in-depth interviews with the case study company's top managers and corresponding qualitative analysis, the top management's requirements for such an implementation could be determined, and through subsequent focus group discussions with a larger group of senior managers, the related topics for a later quantitative survey of the employees and the corresponding indicators for measuring the performance of the implemented process could be developed. By employing a questionnaire, the employees of the case study company and their supply chain partners could be surveyed about the required subjects for a later Supply Chain Risk Management training programme. Furthermore, the indicators for measuring success after implementing the cooperative Supply Chain Risk Management approach at the case study company and its supply chain partners are also analysed quantitatively. The quantitative data gained can then be statistically analysed and provides a basis for testing prespecified hypotheses and explaining causal relationships.

The content-related part of the recommendations for future research is based on this project's research context, which considered only Hidden Champions from the mechanical engineering industry in the cross-border area of the DACH region. Future researchers could (1) extend their studies beyond the DACH region's cross-border area to include other Hidden Champions clusters, such as the Hohenlohe district in the north-eastern part of the state of Baden-Württemberg (Venohr and Lang 2014). Also, the Swiss canton Basle-Country, centred between Germany, France and Switzerland, with the highest per capita concentration of Hidden Champions of Switzerland (see sect. 2.3.2), is a notable example. By expanding the geographical scope of their studies, future researchers could investigate the Hidden Champions in these areas handling Supply Chain Risk Management in cooperation with their supply chain partners. Furthermore, (2) the scope of this study could be extended to other small and medium-sized businesses, not limited exclusively to Hidden Champions, that have internationalised their supply chains to the same extent as the Hidden Champions studied in this research project. In this way, researchers could investigate how companies that are not market-leading have developed their Supply Chain Risk Management processes and, in that way, investigate further how this influences business performance and thus market leadership. Finally, (3) further studies on Supply Chain Risk Management would benefit from investigating in topic-related talent management or the adjacent topic area of Supply Chain Risk Management knowledge management. This investigation could help to identify the competencies required with current and future employees and how these competencies are developed to master frequently volatile markets and a changing professional environment driven by increasing digitalisation.

References

Act on Corporate Due Diligence Obligations in Supply Chains, 2021. Available from: https://www.bmas.de/EN/Services/Press/recent-publications/2021/act-on-corporate-due-diligence-in-supply-chains.html [Accessed 16 September 2022].

Ahern, K., 2014. *Gatekeepers: People who can (and do) stop your research in its tracks* [online]. London: Sage. Available from: https://methods.sagepub.com/case/gatekeepers-people-who-can-and-do-stop-your-research-in-its-tracks [Accessed 30 January 2022].

Al Smadi, S., 2009. Kaizen strategy and the drive for competitiveness: challenges and opportunities. *Competitiveness Review: An International Business Journal* [online], 19 (3), 203–211.

Al-Doori, J. A., 2019. The impact of supply chain collaboration on performance in automotive industry: Empirical evidence. *Journal of Industrial Engineering and Management* [online], 12 (2), 241–253.

Andrade, A. D., 2009. Interpretive research aiming at theory building: Adopting and adapting the case study design. *The Qualitative Report* [online], 14 (1), 42–60.

Andreeva, E. L., Simon, H., Karkh, D. A. and Glukhikh, P. L., 2016. Innovative Entrepreneurship; a Source of Economic Growth in the Region. *Economy of Region* [online], 12 (3), 899–910.

Ataseven, C., Prajogo, D. I. and Nair, A., 2014. ISO 9000 Internalization and Organizational Commitment—Implications for Process Improvement and Operational Performance. *IEEE Transactions on Engineering Management* [online], 61 (1), 5–17.

Ates, A. and Bititci, U., 2011. Change process: a key enabler for building resilient SMEs. *International Journal of Production Research* [online], 49 (18), 5601–5618.

Audretsch, D. B., Lehmann, E. E. and Schenkenhofer, J., 2018. Internationalization strategies of hidden champions: lessons from Germany. *Multinational Business Review* [online], 26 (1), 2–24.

Austrian Institute for SME Research, 2022. *SME Data* [online]. Vienna: Austrian Institute for SME Research. Available from: https://www.kmuforschung.ac.at/facts-and-figures/kmu-daten/?lang=en [Accessed 30 January 2022].

Autry, C. and Sanders, N., 2009. Supply Chain Security: A Dynamic Capabilities Approach. *In:* Zsidisin, G. A. and Ritchie, B., eds. *Supply Chain Risk: A Handbook of Assessment, Management, and Performance* [online]. Boston: Springer, 307–329.

© The Editor(s) (if applicable) and The Author(s), under exclusive license to Springer Fachmedien Wiesbaden GmbH, part of Springer Nature 2026

M. Mayer, *Spanning Organisational and Geographical Boundaries*,
https://doi.org/10.1007/978-3-658-50453-3

Ayyagari, M., Beck, T. and Demirguc-Kunt, A., 2007. Small and Medium Enterprises Across the Globe. *Small Business Economics* [online], 29 (4), 415–434.

Bals, C. and Bals, L., 2019. Exploring fragmentation in the supply chain finance: ecosystem. *In:* Tate, W. L., Bals, L. and Ellram, L. M., eds. *Supply Chain Finance: Risk Management, Resilience and Supplier Management.* London: Kogan Page, 207–224.

Bals, L., Schulze, H., Kelly, S. and Stek, K., 2019. Purchasing and supply management (PSM) competencies: Current and future requirements. *Journal of Purchasing and Supply Management* [online], 25 (5), 100572.

Bals, L. and Turkulainen, V., 2021. Integration of the buyer–supplier interface for Global sourcing. *Operations Management Research* [online], 14 (3), 293–317.

Baptista Nunes, M., Annansingh, F., Eaglestone, B. and Wakefield, R., 2006. Knowledge management issues in knowledge-intensive SMEs. *Journal of Documentation* [online], 62 (1), 101–119.

Barney, J., 1991. Firm Resources and Sustained Competitive Advantage. *Journal of Management*, 17 (1), 99–120.

Beardwell, J. and Thompson, A., 2017. *Human resource management: a contemporary approach* [online]. 8th edition. Harlow: Pearson Education.

Beer, R., Ahn, H.-S. and Leider, S., 2018. *Incentive and competition effects of supplier awards* [online]. Ann Arbor: University of Michigan. Report Number1368.

Berrone, P., Cruz, C. and Gomez-Mejia, L. R., 2012. Socioemotional Wealth in Family Firms: Theoretical Dimensions, Assessment Approaches, and Agenda for Future Research. *Family Business Review* [online], 25 (3), 258–279.

Bessant, J., Kaplinsky, R. and Lamming, R., 2003. Putting supply chain learning into practice. *International Journal of Operations & Production Management* [online], 23 (2), 167–184.

Blackhurst, J., Dunn, K. S. and Craighead, C. W., 2011. An Empirically Derived Framework of Global Supply Resiliency. *Journal of Business Logistics* [online], 32 (4), 374–391.

Bliem, W., Petanovitsch, A. and Schmid, K., 2016. *Dual Vocational Education and Training in Austria, Germany, Liechtenstein and Switzerland* [online]. Vienna: Institut für Bildungsforschung der Wirtschaft.

Blome, C. and Henke, M., 2009. Single Versus Multiple Sourcing: A Supply Risk Management Perspective. *In:* Zsidisin, G. A. and Ritchie, B., eds. *Supply Chain Risk: A Handbook of Assessment, Management, and Performance* [online]. Boston: Springer, 125–135.

Blome, C. and Schoenherr, T., 2011. Supply chain risk management in financial crises—A multiple case-study approach. *International Journal of Production Economics* [online], 134 (1), 43–57.

Bournemouth University Research Ethics Committee (REC), 2021. *8B—Research Ethics Code of Practice: Policy and Procedure* [online]. Bournemouth: Bournemouth University. Version Number: 7.

Bradburn, N. M., Sudman, S. and Wansink, B., 2004. *Asking questions: the definitive guide to questionnaire design—For market research, political polls, and social and health questionnaires* [online]. Rev. edition. San Francisco: Jossey-Bass.

Brannick, T. and Coghlan, D., 2007. In Defense of Being “Native”: The Case for Insider Academic Research. *Organizational Research Methods* [online], 10 (1), 59–74.

Britzelmaier, B., Häberle, M. and Landwehr, M., 2015. Risk management in German small and medium—sized enterprises. *International Journal of Entrepreneurship and Small Business [online]*, 24 (4), 548–571.

Bromiley, P. and Rau, D., 2014. Towards a practice-based view of strategy. *Strategic Management Journal*, 35 (8), 1249–1256.

Bromiley, P. and Rau, D., 2015. Operations management and the resource based view: Another view. *Journal of Operations Management*, 41, 95–106.

Brüning, M. K., 2019. Collaborative recovery from supply chain disruptions. *In:* Bode, C., Bogaschewsky, R., Eßig, M., Lasch, R. and Stölzle, W., eds. *Supply Management Research* [online]. Wiesbaden: Springer, 51–75.

Bryman, A. and Bell, E., 2015. *Business Research Methods* [online]. 4th edition. Oxford: Oxford University Press.

Burkhardt, J., 2019. Specific investments in the supply chain—a literature review on the state-of-the-art knowledge with an outlook on safeguarding mechanisms and avenues for future research. *In:* Bode, C., Bogaschewsky, R., Eßig, M., Lasch, R. and Stölzle, W., eds. *Supply Management Research* [online]. Wiesbaden: Springer, 129–147.

Burrell, G. and Morgan, G.,1982. *Sociological Paradigms and Organisational Analysis.* London: Heinemann.

Buse, S. and Tiwari, R., 2014. *Global innovation strategies of German hidden champions in key emerging markets* [online]. Hamburg: Institute for Technology and Innovation Management (TIM). Report Number 85.

Busse, C. and Mayr, S., 2019. 100 Jahre Trigema – Willkommen bei den Grupps (English title: 100 years of Trigema—Welcome to the Grupps). *Süddeutsche Zeitung* [online], 26. October 2019. Available from: https://www.sueddeutsche.de/wirtschaft/100-jahre-trigema-willkommen-bei-den-grupps-1.4654397?reduced=true [Accessed 30 January 2022].

Cao, M. and Zhang, Q., 2013. *Supply chain collaboration: Roles of Interorganizational Systems, Trust and Collaborative Culture* [online]. London: Springer.

Cappelli, P., 2008. Talent management for the twenty-first century. *Harvard Business Review* [online], 86 (3), 74–81.

Cardoso, S. R., Paula Barbosa-Póvoa, A., Relvas, S. and Novais, A. Q., 2015. Resilience metrics in the assessment of complex supply-chains performance operating under demand uncertainty. *Omega [online]*, 56, 53–73.

Carter, C. R., Kosmol, T. and Kaufmann, L., 2017. Toward a Supply Chain Practice View. *Journal of Supply Chain Management* [online], 53 (1), 114–122.

Carter, C. R., Rogers, D. S. and Choi, T. Y., 2015. Toward the Theory of the Supply Chain. *Journal of Supply Chain Management* [online], 51 (2), 89–97.

Cencic, J., 2019. *Hidden Champions: Austria's Secret Global Market Leaders* [online]. Vienna: Austrian Business Agency. Available from: https://investinaustria.at/en/blog/2019/12/hidden-champions-in-austria.php [Accessed 30 January 2022]

Chen, P.-Y., Chen, K.-Y. and Wu, L.-Y., 2017. The impact of trust and commitment on value creation in asymmetric buyer–seller relationships: the mediation effect of specific asset investments. *Journal of Business & Industrial Marketing* [online], 32 (3), 457–471.

Chiarini, A., 2017. Risk-based thinking according to ISO 9001:2015 standard and the risk sources European manufacturing SMEs intend to manage. *The TQM Journal* [online] 29 (2), 310–323.

Chopra, S. and Sodhi, M. S., 2004. Managing Risk To Avoid Supply-Chain Breakdown. *MIT Sloan Management Review* [online], 46 (1), 53–62.

Christopher, M. and Lee, H., 2004. Mitigating supply chain risk through improved confidence. *International Journal of Physical Distribution & Logistics Management* [online], 34 (5), 388–396.

Christopher, M., Mena, C., Khan, O. and Yurt, O., 2011. Approaches to managing global sourcing risk. *Supply Chain Management: An International Journal* [online], 16 (2), 67–81.

Christopher, M. and Peck, H., 2004. Building the resilient supply chain. *The International Journal of Logistics Management* [online], 15 (2), 1–14.

Christopher, M., Peck, H., Abley, J., Haywood, M., Saw, R., Rutherford, C. and Strathern, M., 2003. *Creating resilient supply chains: a practical guide* [online]. Cranfield: Cranfield University.

Colicchia, C. and Strozzi, F., 2012. Supply chain risk management: a new methodology for a systematic literature review. *Supply Chain Management: An International Journal* [online], 17 (4), 403–418.

Collard, S. and van Teijlingen, E., 2016. Internet-based focus groups: New approaches to an 'old' research method. *Health Prospect* [online], 15 (3), 4–7.

Cooper, M. C. and Ellram, L. M., 1993. Characteristics of supply chain management and the implications for purchasing and logistics strategy. *The International Journal of Logistics Management* [online], 4 (2), 13–24.

Crane, B. and Hartwell, C. J., 2019. Global talent management: A life cycle view of the interaction between human and social capital. *Journal of World Business* [online], 54 (2), 82–92.

Creswell, J. W., 2014. *Research Design: Qualitative, Quantitative, and Mixed Methods Approaches*. 4th edition. Thousand Oaks: Sage.

Credit Suisse AG, 2018. *2018 Swiss SME economy: Successful amid the global competition* [online]. Zurich: Credit Suisse AG.

Curkovic, S., Scannell, T., Wagner, B. and Vitek, M., 2013. Supply chain risk management within the context of COSO's enterprise risk management framework. *Journal of Business Administration Research* [online], 2 (1), 15.

Darby, J. L., Fugate, B. S. and Murray, J. B., 2019. Interpretive research: A complementary approach to seeking knowledge in supply chain management. *The International Journal of Logistics Management* [online], 30 (2), 395–413.

de Oliveira, U. R., Marins, F. A. S., Rocha, H. M. and Salomon, V. A. P., 2017. The ISO 31000 standard in supply chain risk management. *Journal of Cleaner Production* [online], 151, 616–633.

Diehl, D. and Spinler, S., 2013. Defining a common ground for supply chain risk management—a case study in the fast-moving consumer goods industry. *International Journal of Logistics Research and Applications* [online], 16 (4), 311–327.

Din, F.-U., Dolles, H. and Middel, R., 2013. Strategies for small and medium-sized enterprises to compete successfully on the world market: Cases of Swedish hidden champions. *Asian Business & Management* [online], 12 (5), 591–612.

Dolles, H., 2010. Developing world market-leading companies—innovation governance in German small and medium-sized enterprises. *International Journal of Business Environment* [online], 3 (2), 139–158.

Dunne, C., 2011. The place of the literature review in grounded theory research. *International Journal of Social Research Methodology* [online], 14 (2), 111–124.

Dyer, J. and Singh, H., 1998. The Relational View: Cooperative Strategy and Sources of Interorganizational Competitive Advantage. *The Academy of Management Review*, 23.

Eckes, G., 2001. *The Six Sigma Revolution: How General Electric and Others Turned Process Into Profits*. New York: Wiley.

Edwards, F. R. and Canter, M. S., 1995. The collapse of Metallgesellschaft: Unhedgeable risks, poor hedging strategy, or just bad luck? *Journal of Futures Markets*, 15 (3), 211–264.

Eisenhardt, K. M., 1989. Building Theories from Case Study Research. *The Academy of Management Review*, 14 (4), 532–550.

Eisenhardt, K. M., Graebner, M. E. and Sonenshein, S., 2016. Grand Challenges and Inductive Methods: Rigor without Rigor Mortis. *Academy of Management Journal* [online], 59 (4), 1113–1123.

Elahi, E., 2013. Risk management: the next source of competitive advantage. *Foresight* [online], 15 (2), 117–131.

Ellegaard, C., 2008. Supply risk management in a small company perspective. *Supply Chain Management* [online], 13 (6), 425–434.

Ellram, L. M., 1993. A Framework for Total Cost of Ownership. *The International Journal of Logistics Management* [online], 4 (2), 49–60.

Ermann, U., Lang, T. and Megerle, M., 2011. Weltmarktführer: ein räumlicher und zeitlicher Überblick (English title: World market leaders: a geographical and chronological overview). *Nationalatlas aktuell* [online]. 5 (11.2011). Available from: http://aktuell.nationalatlas.de/weltmarktfuehrer-11_11-2011-0-html/ [Accessed 30 January 2022].

European Commission, 2022. Just and sustainable economy: Commission lays down rules for companies to respect human rights and environment in global value chains [online]. Brussels: European Commission. Available from: https://ec.europa.eu/commission/presscorner/detail/en/ip_22_1145 [Accessed 16 September 2022].

Ezzy, D., 2002. *Qualitative analysis: practice and innovation* [online]. Crows Nest: Allen & Unwin.

Faisal, M. N., Banwet, D. and Shankar, R., 2007. Supply chain risk management in SMEs: analysing the barriers. *International Journal of Management and Enterprise Development*, 4 (5), 588–607.

Falkner, E. M. and Hiebl, M. R. W., 2015. Risk management in SMEs: a systematic review of available evidence. *Journal of Risk Finance* [online], 16 (2), 122- 144.

Farahani, R. Z., Rezapour, S., Drezner, T. and Fallah, S., 2014. Competitive supply chain network design: An overview of classifications, models, solution techniques and applications. *Omega* [online], 45, 92–118.

Farris, T. and Hightower, S., 2019. Supply chain finance opportunities by managing payables discounts. *In:* Tate, W. L., Bals, L. and Ellram, L. M., eds. *Supply Chain Finance: Risk Management, Resilience and Supplier Management.* London: Kogan Page, 123–137.

Fawcett, S. E. and Magnan, G. M., 2002. The rhetoric and reality of supply chain integration. *International Journal of Physical Distribution & Logistics Management* [online], 32 (5), 339–361.

Felden, B., Hack, A. and Hoon, C., 2019. *Management von Familienunternehmen* (English title: *Management of Family Businesses*) [online]. 2nd edition. Wiesbaden: Springer Gabler.

Finch, P., 2004. Supply chain risk management. *Supply Chain Management* [online], 9 (2), 183–196.

Flick, U., 2018. *An introduction to qualitative research* [online]. 6th edition. London: Sage.

FMTI, 2021. *Metaltechnology Austria—Facts and Figures 2021* [online]. Vienna: Metaltechnology Austria. Available from: https://www.metalltechnischeindustrie.at/fileadmin/content/Dokumente/Zahlen_Daten_Fakten/Englisch/facts_and_figures_2021.pdf [Accessed 30 January 2022]

Friday, D., Ryan, S., Sridharan, R. and Collins, D., 2018. Collaborative risk management: a systematic literature review. *International Journal of Physical Distribution & Logistics Management* [online], 48 (3), 231–253.

Garaus, C., Güttel, W. H., Konlechner, S., Koprax, I., Lackner, H., Link, K. and Müller, B., 2016. Bridging knowledge in ambidextrous HRM systems: empirical evidence from hidden champions. *The International Journal of Human Resource Management* [online], 27 (3), 355–381.

Gebauer, H., Fischer, T. and Fleisch, E., 2009. Management policies of Swiss product-manufacturing companies: critical factors in international competition (Part 1). *Business Strategy Series*, 10 (3), 111–120.

Gehman, J., Glaser, V. L., Eisenhardt, K. M., Gioia, D., Langley, A. and Corley, K. G., 2018. Finding Theory–Method Fit: A Comparison of Three Qualitative Approaches to Theory Building. *Journal of Management Inquiry* [online], 27 (3), 284–300.

Gill, P., Stewart, K., Treasure, E. and Chadwick, B., 2008. Methods of data collection in qualitative research: interviews and focus groups. *British Dental Journal* [online], 204, 291–295.

Gioia, D. A., Corley, K. G. and Hamilton, A. L., 2013. Seeking Qualitative Rigor in Inductive Research: Notes on the Gioia Methodology. *Organizational Research Methods* [online], 16 (1), 15–31.

Gioia, D. A., Thomas, J. B., Clark, S. M. and Chittipeddi, K., 1994. Symbolism and Strategic Change in Academia: The Dynamics of Sensemaking and Influence. *Organization Science* [online], 5 (3), 363–383.

Glas, A. H. and Kleemann, F. C., 2016. The Impact of Industry 4.0 on Procurement and Supply Management: A Conceptual and Qualitative Analysis. *International Journal of Business and Management Invention* [online], 5 (6), 55–66.

Golgeci, I. and Ponomarov, S. Y., 2013. Does firm innovativeness enable effective responses to supply chain disruptions? An empirical study. *Supply Chain Management: An International Journal* [online], 18 (6), 604–617.

Gómez-Mejía, L. R., Haynes, K. T., Núñez-Nickel, M., Jacobson, K. J. L. and Moyano-Fuentes, J., 2007. Socioemotional Wealth and Business Risks in Family-controlled Firms: Evidence from Spanish Olive Oil Mills. *Administrative Science Quarterly* [online], 52 (1), 106–137.

Gredal, P., Panyi, Z., Kinra, A. and Kotzab, H., 2017. What Hinders the Implementation of the Supply Chain Risk Management Process into Practice Organizations? *In:* Freitag, M.,

Kotzab, H. and Pannek, J., eds. *Dynamics in Logistics: Proceedings of the 5th International Conference LDIC, 2016, Bremen, Germany, 22–25 February 2016.* [online]. Cham: Springer, 151–161.

Gruenwald, R. K., 2013. Entrepreneurship Challenges in High-Growth Companies and Consequences for SME Policy. *Entrepreneurial Business and Economics Review* [online], 1 (1), 41–54.

Güller, M. and Henke, M., 2019. Resilience Assessment in Complex Supply Networks. *In:* Zsidisin, G. A. and Henke, M., eds. *Revisiting Supply Chain Risk* [online]. Cham: Springer, 73–98.

Hail, C., Hurst, B. and Camp, D., 2011. Peer Debriefing: Teachers' Reflective Practices for Professional Growth. *Critical Questions in Education* [online], 2 (2), 74–83.

Hallikas, J. and Lintukangas, K., 2016. Purchasing and supply: An investigation of risk management performance. *International Journal of Production Economics* [online], 171, 487–494.

Hallikas, J., Lintukangas, K. and Grudinschi, D., 2019. Sustainability Risk Management in Supply Chain. *In:* Zsidisin, G. A. and Henke, M., eds. 2019. *Revisiting Supply Chain Risk* [online]. Cham: Springer, 265–278.

Hauswald, H., Hack, A., Kellermanns, F. W. and Patzelt, H., 2016. Attracting New Talent to Family Firms: Who is Attracted and under what Conditions? *Entrepreneurship Theory and Practice* [online], 40 (5), 963–989.

Heckmann, I., Comes, T. and Nickel, S., 2015. A critical review on supply chain risk—Definition, measure and modeling. *Omega* [online], 52, 119–132.

Helfferich, C., 2014. Leitfaden- und Experteninterviews (English title: Guided and Expert Interviews). *In:* Baur, N. and Blasius, J., eds. *Handbuch Methoden der empirischen Sozialforschung* [online]. Wiesbaden: Springer, 559–574.

Helmold, M., 2014. *Establishing a best practice model of supplier relationship management (SRM) for multinational manufacturing companies in the European transportation industry* [online]. Thesis (DBA). University of Gloucestershire, Available from: https://eprints.glos.ac.uk/632/1/M%20%20Helmold_DBA%20Thesis.pdf [Accessed 30 January 2022].

Ho, W., Zheng, T., Yildiz, H. and Talluri, S., 2015. Supply chain risk management: a literature review. *International Journal of Production Research* [online], 53 (16), 5031–5069.

Hofmann, E., 2005. Supply chain finance: some conceptual insights [online]. *In:* Lasch, R. and Janker, C. G., eds. *Logistik Management: Innovative Logistikkonzepte 2005,* Dresden *04–06 October 2005. Wiesbaden:* Deutscher Universitätsverlag. 203–214. Available from: https://www.alexandria.unisg.ch/22597/1/Supply%20Chain%20Finance.pdf [Accessed 30 January 2022].

Hofmann, E., 2011. Natural hedging as a risk prophylaxis and supplier financing instrument in automotive supply chains. *Supply Chain Management* [online], 16 (2), 128–141.

Hofmann, E. and Staiger, F., 2020. *Beschaffungskompetenzen 4.0: Berufsbilder im Zeitalter des digitalisierten Einkaufs* (English title: *Sourcing competencies 4.0: Career profiles in the age of digitalised purchasing*) [online]. Berlin: Springer Gabler.

Hohenstein, N.-O., Feisel, E., Hartmann, E. and Giunipero, L., 2014. Research on the phenomenon of supply chain resilience: A systematic review and paths for further investigation. *International Journal of Physical Distribution & Logistics Management* [online], 45 (1/2), 90–117.

Huber, A. S. and Gith, R., 2009. Equity-Rating als Risk-Performance-Instrument (English title: Equity-Rating as a Risk-Performance-Instrument). *In:* Hilz-Ward, R. M. and Everling, O., eds. *Risk Performance Management: Chancen für ein besseres Rating* [online]. Wiesbaden: Gabler, 111–131.

Institut für Mittelstandsforschung Bonn, 2022. *Macro-economic significance of SMEs* [online]. Bonn: Institut für Mittelstandsforschung. Available from: https://www.ifm-bonn.org/en/statistics/overview-mittelstand/macro-economic-significance-of-smes/deutschland [Accessed 30 January 2022].

International Organization for Standardization, 2015. *9001:2015 Quality Management Systems—Requirements* [online]. Geneva: International Standardization Organization.

International Organization for Standardization, 2018. *ISO 31000:2018 Risk Management—Guidelines* [online]. Geneva: International Standardization Organization.

Ivanov, D., Sokolov, B. and Dolgui, A., 2014. The Ripple effect in supply chains: trade-off 'efficiency-flexibility-resilience' in disruption management. *International Journal of Production Research* [online], 52 (7), 2154–2172.

Ivanov, D., 2020. Viable supply chain model: integrating agility, resilience and sustainability perspectives—lessons from and thinking beyond the COVID-19 pandemic. *Annals of Operations Research* [online].

Ivanov, D. and Dolgui, A., 2020. Viability of intertwined supply networks: extending the supply chain resilience angles towards survivability. A position paper motivated by COVID-19 outbreak. *International Journal of Production Research* [online], 58 (10), 2904–2915.

Ivanov, D., 2021. Supply Chain Viability and the COVID-19 pandemic: a conceptual and formal generalisation of four major adaptation strategies. *International Journal of Production Research* [online], 59 (12), 3535–3552.

Ivanov, D., Tsipoulanidis, A. and Schönberger, J., 2021. Processes, Systems, and Models. In: Ivanov, D., Tsipoulanidis, A. and Schönberger, J., eds. Global Supply Chain and Operations Management: A Decision-Oriented Introduction to the Creation of Value [online]. Cham: Springer, 49–83.

Jacob, F. and Strube, G., 2008. Why Go Global? The Multinational Imperative. *In:* Abele, E., Meyer, T., Näher, U., Strube, G. and Sykes, R., eds. *Global Production: A Handbook for Strategy and Implementation* [online]. Berlin: Springer, 2–33.

Jaine, T., 2014. Elevenses. *In:* The Oxford Companion to Food [online]. 3rd Edition. Oxford. Oxford University Press.

Jayaram, J., Dixit, M. and Motwani, J., 2014. Supply chain management capability of small and medium sized family businesses in India: A multiple case study approach. *International Journal of Production Economics* [online], 147 (Part B), 472–485.

Jüttner, U., 2005. Supply chain risk management: Understanding the business requirements from a practitioner perspective. *The International Journal of Logistics Management* [online], 16 (1), 120–141.

Jüttner, U., Peck, H. and Christopher, M., 2003. Supply chain risk management: outlining an agenda for future research. *International Journal of Logistics: Research and Applications* [online], 6 (4), 197–210.

Kache, F. and Seuring, S., 2014. Linking collaboration and integration to risk and performance in supply chains via a review of literature reviews. *Supply Chain Management: An International Journal* [online], 19 (5/6), 664–682.

Kähkönen, A.-K., 2014. The influence of power position on the depth of collaboration. *Supply Chain Management: An International Journal* [online], 19 (1), 17–30.

Karjalainen, K., Kemppainen, K. and van Raaij, E. M., 2008. Non-Compliant Work Behaviour in Purchasing: An Exploration of Reasons Behind Maverick Buying. *Journal of Business Ethics* [online], 85 (2), 245.

Kaufmann, L., Esslinger, J. and Carter, C. R., 2018. Toward Relationship Resilience: Managing Buyer-Induced Breaches of Psychological Contracts During Joint Buyer–Supplier Projects. *Journal of Supply Chain Management* [online], 54 (4), 62–85.

Kern, D., Moser, R., Sundaresan, N. and Hartmann, E., 2011. Purchasing Competence: A Stakeholder-Based Framework for Chief Purchasing Officers. *Journal of Business Logistics* [online],32 (2), 122–138.

Kilubi, I. and Haasis, H. D., 2015. Supply chain risk management enablers—A framework development through systematic review of the literature from 2000 to 2015. *International Journal of Business Science and Applied Management* [online], 10 (1), 35–54.

Kirchoff, J. F., Omar, A. and Fugate, B. S., 2016. A Behavioral Theory of Sustainable Supply Chain Management Decision Making in Non-exemplar Firms. *Journal of Supply Chain Management* [online], 52 (1), 41–65.

Kitzinger, J., 2006. Focus Groups. *In:* Pope, C. and Mays, N., eds. *Qualitative Research in Health Care* [online]. Oxford: Blackwell Publishing Ltd., 21–31.

Kivunja, C., 2018. Distinguishing between Theory, Theoretical Framework, and Conceptual Framework: A Systematic Review of Lessons from the Field. *International Journal of Higher Education* [online], 7 (6), 44–53.

Kleindorfer, P. R. and Saad, G. H., 2005. Managing Disruption Risks in Supply Chains. *Production and Operations Management* [online], 14 (1), 53–68.

Kosmol, T., Reimann, F. and Kaufmann, L., 2019. You'll never walk alone: Why we need a supply chain practice view on digital procurement. *Journal of Purchasing and Supply Management* [online], 25 (4), 100553.

Kulp, S. C., Lee, H. L. and Ofek, E., 2004. Manufacturer Benefits from Information Integration with Retail Customers. *Management Science* [online], 50 (4), 431–444.

Kumar, S. and Anbanandam, R., 2020. Impact of risk management culture on supply chain resilience: An empirical study from Indian manufacturing industry. *Proceedings of the Institution of Mechanical Engineers, Part O: Journal of Risk and Reliability* [online], 234 (2), 246–259.

Landau, C., Karna, A. and Täube, F., 2016. The Internationalization of Hidden Champions. *In:* Merchant, H., ed. *Handbook of Contemporary Research on Emerging Markets* [online]. Cheltenham: Edward Elgar Publishing, 292–316.

Latour, A., 2001. A Fire in Albuquerque Sparks Crisis For European Cell-Phone Giants. The *Wall Street Journal* [online], *29 January 2001 Available from:* https://www.wsj.com/articles/SB980720939804883010 [Accessed 30 January 2022].

Lau, A. K. W., Tang, E. and Yam, R. C. M., 2010. Effects of Supplier and Customer Integration on Product Innovation and Performance: Empirical Evidence in Hong Kong Manufacturers. *Journal of Product Innovation Management* [online], 27 (5), 761–777.

Le Dain, M.-A., Merminod, V. and Yager, M., 2020. Collaborative practices in new product development projects involving suppliers. *Production Planning & Control* [online], 31 (4), 308–321.

Lebedev, P., 2013. Financial Aspects of HCs Business Models. *In:* McKiernan, P. and Purg, D., eds. *Hidden Champions in CEE and Turkey: Carving Out a Global Niche* [online]. Berlin: Springer, 95–107.

Lee, H. L., Padmanabhan, V. and Whang, S., 1997. Information Distortion in a Supply Chain: The Bullwhip Effect. *Management Science* [online], 43 (4), 546–558.

Lee, J. and Wang, J., 2017. Developing entrepreneurial resilience: implications for human resource development. *European Journal of Training and Development* [online], 41 (6), 519–539.

Lehrer, M. and Schmid, S., 2015. Germany's industrial family firms: Prospering islands of social capital in a financialized world? *Competition & Change* [online], 19 (4), 301–316.

Leisy, B. and Pyron, D., 2009. Talent Management Takes On New Urgency. *Compensation & Benefits Review*[online], 41 (4), 58–63.

Lekkakos, S. D. and Serrano, A., 2016. Supply chain finance for small and medium sized enterprises: the case of reverse factoring. *International Journal of Physical Distribution & Logistics Management* [online], 46 (4), 367–392.

Leon, A. C., Davis, L. L. and Kraemer, H. C., 2011. The Role and Interpretation of Pilot Studies in Clinical Research. *Journal of Psychiatric Research* [online], 45 (5), 626–629.

Levitt, T., 1983. The globalization of markets. *Harvard Business Review* [online], 61 (3), 92–102.

Lewis, R. E. and Heckman, R. J., 2006. Talent management: A critical review. *Human Resource Management Review* [online], 16 (2), 139–154.

Li, G., Fan, H., Lee, P. K. C. and Cheng, T. C. E., 2015. Joint supply chain risk management: An agency and collaboration perspective. *International Journal of Production Economics* [online], 164, 83–94.

Lincoln, Y. S. and Guba, E. G., 1985. *Naturalistic Inquiry*. Newbury Park: Sage.

Liu, C.-L., Shang, K.-C., Lirn, T.-C., Lai, K.-H. and Lun, Y. H. V., 2018. Supply chain resilience, firm performance, and management policies in the liner shipping industry. *Transportation Research Part A: Policy and Practice* [online], 110, 202–219.

Manuj, I. and Mentzer, J. T., 2008. Global supply chain risk management strategies. *International Journal of Physical Distribution & Logistics Management* [online] 38 (3), 192–223.

Maier, S. L. and Monahan, B. A., 2009. How Close Is Too Close? Balancing Closeness And Detachment In Qualitative Research. *Deviant Behavior* [online], 31 (1), 1–32.

Martínez-Romero, M. J. and Rojo-Ramírez, A. A., 2016. SEW: Looking for a definition and controversial issues. *European Journal of Family Business*, 6 (1), 1–9.

Mayer, J. A., Borchardt, M. and Pereira, G. M., 2016. Methodology for the collaboration in supply chains with a focus on continuous improvement. *Ingeniería e Investigación* [online], 36, 51–59.

McLafferty, I., 2004. Focus group interviews as a data collecting strategy. *Journal of Advanced Nursing* [online], 48 (2), 187–194.

McLean, T., 2017. *On Time, in Full: Achieving Perfect Delivery with Lean Thinking in Purchasing, Supply Chain, and Production Planning* [online]. Boca Raton: CRC Press.

McNamara, C., 1998. Founder's Syndrome: How Founders and Their Organizations Recover. *Nonprofit World* [online], 16 (6), 38–41.

Mentzer, J. T. and Firman, J., 1994. Logistics Control Systems in the 21st Century. *Journal of Business Logistics* [online], 15 (1), 215–228.

Meuser, M. and Nagel, U., 2009. Das Experteninterview – konzeptionelle Grundlagen und methodische Anlage (English title: The expert interview—conceptual foundations and methodological design). *In:* Pickel, S., Pickel, G., Lauth, H.-J. and Jahn, D., eds. *Methoden der vergleichenden Politik- und Sozialwissenschaft* [online]. Wiesbaden: VS Verlag für Sozialwissenschaften, 465–479.

Merrilees, B., Blackburn, J., Tiessen, J.H. and Lindman, M., 2001. Hidden (SME) champions: the role of innovation and strategy [online]. *In:* Chetty, S. and Collins, B., eds. *Australian and New Zealand Marketing Academy Conference*, Auckland 1–5 December 2001. Auckland, NZ: College of Business, Massey University. 1–5. Available from: http://citeseerx.ist.psu.edu/viewdoc/download?doi=10.1.1.195.1825&rep=rep1&type=pdf [Accessed 30 January 2022].

Meyer-Stamer, J. and Wältring, J., 2000. *Behind the myth of the Mittelstand economy: the institutional environment supporting small and medium-sized enterprises in Germany* [online]. Duisburg: Institute for Development and Peace. Report Number 46/2000.

Michaels, A. and Grüning, M., 2017. Relationship of corporate social responsibility disclosure on information asymmetry and the cost of capital. *Journal of Management Control* [online], 28 (3), 251–274.

Mietzner, M., Proelss, J. and Schweizer, D., 2018. Hidden champions or black sheep? The role of underpricing in the German mini-bond market. *Small Business Economics* [online], 50 (2), 375–395.

Miller, D. and Le Breton–Miller, I., 2014. Deconstructing Socioemotional Wealth. *Entrepreneurship Theory and Practice* [online], 38 (4), 713–720.

Miller, H. E. and Engemann, K. J., 2019. Resilience and Sustainability in Supply Chains. *In:* Zsidisin, G. A. and Henke, M., eds. *Revisiting Supply Chain Risk* [online]. Cham: Springer, 251–263.

Moen, R. and Norman, C., 2009. Evolution of the PDCA cycle [online]. *In:* Suzuki, T., ed. Proceedings of 7th Asian Network for Quality *Congress*, Tokyo 15–18 September 2009. Tokyo: Asian Network for Quality (ANQ). Available from: https://citeseerx.ist.psu.edu/viewdoc/download?doi=10.1.1.470.5465&rep=rep1&type=pdf [Accessed 22 January 2022].

Mohr, G., 2009. *Supply Chain Sourcing* [online]. *Wiesbaden: Gabler.*

Mohr, J. J. and Sengupta, S., 2002. Managing the paradox of inter-firm learning: the role of governance mechanisms. *Journal of Business & Industrial Marketing* [online], 17 (4), 282–301.

Moosa, K. and Sajid, A., 2010. Critical analysis of Six Sigma implementation. *Total Quality Management & Business Excellence* [online], 21 (7), 745–759.

Müller, C., 2019. *Weltmarktführerindex* (English title: *World Market Leaders Index)* [online]. Stuttgart: UEC University Entrepreneurship Cooperation gGmbH. Available from: https://www.weltmarktfuehrerindex.de [Accessed 30 January 2022].

Nalebuff, B. J. and Brandenburger, A. M., 1997. Co-opetition: Competitive and cooperative business strategies for the digital economy. *Strategy & Leadership* [online], 25 (6), 28–33.

Neubauer, R. M., 2011. *Business Models in the Area of Logistics* [online]. Wiesbaden: Gabler.

Norrman, A. and Jansson, U., 2004. Ericsson's proactive supply chain risk management approach after a serious sub-supplier accident. *International Journal of Physical Distribution & Logistics Management* [online], 34 (5), 434–456.

Okongwu, U., Franck, B. and Moncef, B., 2015. Causal linkages between supply chain management practices and performance: A balanced scorecard strategy map perspective. *Journal of Manufacturing Technology Management* [online], 26 (5), 678–702.

Okumus, F., Altinay, L. and Roper, A., 2007. Gaining access for research: Reflections from Experience. *Annals of Tourism Research* [online], 34 (1), 7–26.

OLD, 2022. Mediate. *In: Oxford learner's dictionary* [online]. Oxford: Oxford University Press

Otto, A. and Kotzab, H., 2001. Der Beitrag des Supply Chain Management zum Management von Supply Chains (English title: The contribution of supply chain management to the management of supply chains). *Schmalenbachs Zeitschrift für betriebswirtschaftliche Forschung* [online], 53 (2), 157–176.

Pachl, W. (Ed.). 2016. Maculan Holding AG. In: Austria-Forum [online]. Graz: Austria-Forum.

Pagell, M., Flynn, B. B., Fugate, B. and Cantor, D. E., 2022. Developing purchasing and supply management theory. *In*: Tate, W., Ellram, L. and Bals, L., eds. *Handbook of Theories for Purchasing, Supply Chain and Management Research* [online]. Cheltenham: Edward Elgar Publishing, 29–47.

Pandey, A., 2021. *5 threats to the global economy in 2022* [online]. Bonn: Deutsche Welle. Available from: https://p.dw.com/p/442oR [Accessed 30 January]

Parast, M. M., Sabahi, S. and Kamalahmadi, M., 2019. The Relationship Between Firm Resilience to Supply Chain Disruptions and Firm Innovation. *In:* Zsidisin, G. A. and Henke, M., eds. *Revisiting Supply Chain Risk* [online]. Cham: Springer, 279–298.

Partridge, J., 2021. Global chip shortage hits car production in China and Japan. *The Guardian* [online]. 30 June 2021. Available from: https://www.theguardian.com/world/2021/jun/30/global-semiconductor-shortage-affecting-car-production-in-china-and-japan [Accessed 30 January 2022].

Peteraf, M. A., 1993. The Cornerstones of Competitive Advantage: A Resource-Based View. *Strategic Management Journal* [online], 14 (3), 179–191.

Pfohl, H.-C., Köhler, H. and Thomas, D., 2010. State of the art in supply chain risk management research: empirical and conceptual findings and a roadmap for the implementation in practice. *Logistics Research* [online], 2 (1), 33–44.

Porter, M. E., 1985. The *Competitive Advantage: Creating and Sustaining Superior Performance.* New York: Free Press.

Poulis, K., Poulis, E. and Plakoyiannaki, E., 2013. The role of context in case study selection: An international business perspective. *International Business Review* [online], 22 (1), 304–314.

Pundy, B., 2015. Value Orientation and Contrasting Emphasis on Processes in Organizations A Comparative Analysis in the German-speaking Cultural Cluster. *In:* Covarrubias Venegas, B., Fink, G. and Mayrhofer, W., eds. *IACCM Annual Conference*, Vienna. IACCM and SIETAR Austria. 375. Available from: https://docplayer.net/13027159-Contemporary-approaches-in-training-education-for-cross-cultural-competence-potentials-challenges-and-its-limits-conference-proceedings.html [Accessed 30 January2022].

Ramachandran, V. S. and Blakeslee, S., 2003. *Phantoms in the brain: human nature and the architecture of the mind*. London: Fourth Estate.

Rammer, C. and Frietsch, R., 2015. *Global Champions und Hidden Champions: Internationale Konzerne und KMU im Innovationswettbewerb* (English title: *Global Champions and Hidden Champions: International corporations and SMEs in innovation competition*) [online]. Karlsruhe: Fraunhofer-Institut für System- und Innovationsforschung (ISI). Report Number 45.

Ragatz, G. L., Handfield, R. B. and Scannell, T. V., 1997. Success Factors for Integrating Suppliers into New Product Development. *The Journal of Product Innovation Management* [online], 14 (3), 190–202.

Rao, S. and Goldsby, T. J., 2009. Supply chain risks: a review and typology. *The International Journal of Logistics Management* [online], 20 (1), 97–123.

Ritchie, B. and Brindley, C., 2000. Disintermediation, disintegration and risk in the SME global supply chain. *Management Decision* [online], 38 (8), 575–583.

Romdhane, T. B., Badreddine, A. and Sansa, M., 2017. A new model to implement Six Sigma in small- and medium-sized enterprises. *International Journal of Production Research* [online], 55 (15), 4319–4340.

Romeike, F. and Stallinger, M., 2021. *Stochastische Szenariosimulation in der Unternehmenspraxis* (English title: *Stochastic scenario simulation in business practice*) [online]. Wiesbaden: Springer.

Rumelt, R. P., 1984. Toward a Strategic Theory of the Firm. *In:* Lamb, R. B., ed. *Competitive Strategic Management.* Englewood Cliffs: Prentice-Hall, 556–570.

Saier, M. C., 2017. Going back to the roots of W.A. Shewhart (and further) and introduction of a new CPD cycle. *International Journal of Managing Projects in Business* [online], 10 (1), 143–166.

Sarker, S., 2019. Differentiating Between Supply and Supplier Risk for Better Supply Chain Risk Management. *In:* Zsidisin, G. A. and Henke, M., eds. *Revisiting Supply Chain Risk* [online]. Cham: Springer, 315–328.

Sarkis, J., Cohen, M. J., Dewick, P. and Schröder, P., 2020. A brave new world: Lessons from the COVID-19 pandemic for transitioning to sustainable supply and production. *Resources, Conservation and Recycling* [online], 159, 104894.

Saunders, M. and Lewis, P., 2012. *Doing research in business and management: an essential guide to planning your project*. Harlow: Pearson Education.

Saunders, M., Lewis, P. and Thornhill, A., 2009. *Research methods for business students* [online]. 5th edition. Harlow: Pearson Education.

Schiemann, W. A., 2014. From talent management to talent optimization. *Journal of World Business* [online], 49 (2), 281–288.

Schlepphorst, S., Schlömer-Laufen, N. and Holz, M., 2016. *Determinants of hidden champions—Evidence from Germany* [online]. Bonn: Institut für Mittelstandsforschung. Report Number 03/16

Schlüter, T., 2018. *Performance Management in a German Hidden Champion—an Action Research investigation* [online]. Thesis (DBA). University of Bradford. Available from: https://bradscholars.brad.ac.uk/handle/10454/17397 [Accessed 30 January 2022].

Schlüter, F., 2019. Procedure Model for Supply Chain Digitalization Scenarios for a Data-Driven Supply Chain Risk Management. *In:* Zsidisin, G. A. and Henke, M., eds. *Revisiting Supply Chain Risk* [online]. Cham: Springer, 137–154.

Schmieder, M., 2018. Hidden Champions Benchmarking. *In:* Büchler, J.-P., ed. *Fallstudienkompendium Hidden Champions* [online]. Wiesbaden: Springer 197–222.

Schoenherr, T., Rao Tummala, V. M. and Harrison, T. P., 2008. Assessing supply chain risks with the analytic hierarchy process: Providing decision support for the offshoring decision by a US manufacturing company. *Journal of Purchasing and Supply Management* [online], 14 (2), 100–111.

Schröder, M., 2017. How mature is your supply chain risk management? A self assessment model for supply chain risk management maturity [online]. *In:* Hellström, D., Kembro, J., Bodnar, H., *eds. Nordic Logistics Research Network* (NOFOMA) *2017: The 29th NOFOMA Conference: Taking on grand challenges,* Lund 8–9 June 2017. Lund: Lund University. Available from: https://tore.tuhh.de/bitstream/11420/8753/1/Nofoma_2017_SCRMMaturityModel.pdf [Accessed 30 January 2022]

Schulze, H., Bals, L. and Johnsen, T. E., 2019. Individual competences for sustainable purchasing and supply management (SPSM). *International Journal of Physical Distribution & Logistics Management* [online] 49 (3), 287–304.

Schulze, H. and Bals, L., 2020. Implementing sustainable purchasing and supply management (SPSM): A Delphi study on competences needed by purchasing and supply management (PSM) professionals. *Journal of Purchasing and Supply Management* [online], 26 (4), 100625.

Schumpeter, J. A., 1942. *Capitalism, socialism, and democracy* [online]. Reprint, London: Routledge, 2010.

Schwartz, S., 2006. A theory of cultural value orientations: Explication and applications. *Comparative Sociology* [online], 5 (2–3), 137–182.

Schwass, J. and Glemser, A.-C., 2016. *Wise Family Business: Family Identity Steering Brand Success* [online]. London: Palgrave Macmillan.

Schweiger, J., 2016. Concept of a Purchasing and Supply Management Maturity Framework. *In:* Bogaschewsky, R., Eßig, M., Lasch, R. and Stölzle, W., eds. *Supply Management Research* [online]. Wiesbaden: Springer, 153–176.

Seemiller, C. and Grace, M., 2017. Generation Z: Educating and Engaging the Next Generation of Students. *About Campus* [online], 22 (3), 21–26.

Shaw, S., Grant, D. B. and Mangan, J., 2021. A supply chain practice-based view of enablers, inhibitors and benefits for environmental supply chain performance measurement. *Production Planning & Control* [online], 32 (5), 382–396.

Sheffi, Y., 2019. Preparing for the Worst. *In:* Zsidisin, G. A. and Henke, M., eds. *Revisiting Supply Chain Risk* [online]. Cham: Springer, 155–168.

Sheffi, Y. and Rice, J. B. J., 2005. A supply chain view of the resilient enterprise. *MIT Sloan management review* [online], 47 (1), 41.

Sheffi, Y., Rice, J. J., Fleck, J. and Caniato, F., 2003. A Supply Chain Response to Global Terrorism: A Situation Scan [online]. *In:* Spina, G., Vinelli, A., Cagliano, R., Kalchschmidt, M., Romano, P. and Salvador, F., eds. *One world? One view of OM? The challenges of integrating research & practice. Proceedings of the first EurOMA and POMS joint Conference,* Cernobbio 16–18 June 2003. Padua: SGEditoriali. 291–300, Available from: http://web.mit.edu/scresponse/repository/euroma_paper_041603. doc [Accessed 30 January 2022]

Simon, H., 1990. „Hidden Champions“: Speerspitze der deutschen Wirtschaft (English title: “Hidden Champions”: Spearhead of the German Economy). *Journal of business economics,* 60 (9), 875–890.

Simon, H., 1992. Service Policies of German Manufacturers: Critical Factors in International Competition. *European Management Journal* [online], 10 (4), 404–411.

Simon, H., 1996. *Hidden Champions: lessons from 500 of the world's best unknown companies.* Boston: Harvard Business School Press.

Simon, H., 2009. *Hidden Champions of the Twenty-First Century: The Success Strategies of Unknown World Market Leaders* [online]. New York: Springer.

Simon, H., 2012. Hidden Champions—Aufbruch nach Globalia: Die Erfolgsstrategien unbekannter Weltmarktführer (English title: *Hidden Champions—Setting out for Globalia: the success strategies of unknown world market leaders)* [online]. Frankfurt: Campus Verlag.

Simon, H., 2017. Why Germany Still Has So Many Middle-Class Manufacturing Jobs. *Harvard Business Review* [online]. *02 May 2017. Available from:* https://hbr.org/2017/05/why-germany-still-has-so-many-middle-class-manufacturing-jobs [Accessed 30 January 2022].

Simon, H., 2019. Die digitalen Hidden Champions (English title: The digital Hidden Champions). *Harvard-Business-Manager* [online], *16 November 2020.* Available from: https://www.manager-magazin.de/harvard/digitalisierung/weltmarktfuehrer-aus-dem-mittelstand-die-digitalen-hidden-champions-a-00000000-0002-0001-0000-000166368289 [Accessed 30 January 2022].

Simon, H., 2020a. Erfolgsfaktoren der Hidden Champions – Welche Rolle Strategie, Führung und Digitalisierung spielen (English title: Success factors of the Hidden Champions: What role strategy, leadership and digitalisation play). *Zeitschrift Führung und Organisation [online],* (5), 285–290.

Simon, H., 2020b. Erfolgsfaktoren von Hidden Champions bei der Digitalisierung (English title: Success factors of Hidden Champions during digitalisation). *In:* Bruhn, M., Burmann, C., Kirchgeorg, M., eds. *Marketing Weiterdenken* [online]. Wiesbaden: Springer Gabler.

Simon, H., 2021. *Hidden Champions: Die neuen Spielregeln im chinesischen Jahrhundert (English title: Hidden Champions: The new ground rules in the Chinese century) [online].* Frankfurt: Campus Verlag.

SME Portal, 2019. *Figures on SMEs: Essential Points in Brief* [online]. Bern: Federal Chancellery. Available from: https://www.kmu.admin.ch/kmu/en/home/facts-and-trends/facts-and-figures/figures-smes.html [Accessed 30 January 2022].

Sodhi, M. S., Son, B.-G. and Tang, C. S., 2012. Researchers' Perspectives on Supply Chain Risk Management. *Production and Operations Management* [online], 21 (1), 1–13.

Sodhi, M. S. and Tang, C., 2011. Supply Chain Risk Management. *In:* James J. Cochran, J. J., ed. *Wiley Encyclopedia of Operations Research and Management Science* [online]. John Wiley & Sons

Sodhi, M. S. and Tang, C. S., 2012. *Managing supply chain risk* [online]. New York: Springer.

Sokovic, M., Pavletic, D. and Pipan, K., 2010. Quality improvement methodologies—PDCA cycle, RADAR matrix, DMAIC and DFSS. *Journal of Achievements in Materials and Manufacturing Engineering* [online], 43 (1), 476–483.

Spekman, R. E., Spear, J. and Kamauff, J., 2002. Supply chain competency: learning as a key component. *Supply Chain Management: An International Journal* [online], 7 (1), 41–55.

Spina, G., Caniato, F., Luzzini, D. and Ronchi, S., 2016. Assessing the use of External Grand Theories in Purchasing and Supply Management research. *Journal of Purchasing and Supply Management* [online], 22 (1), 18–30.

Steinle, C. and Schiele, H., 2008. Limits to global sourcing?: Strategic consequences of dependency on international suppliers: Cluster theory, resource-based view and case studies. *Journal of Purchasing and Supply Management* [online], 14 (1), 3–14.

Stevens, G. C., 1989. Integrating the Supply Chain. *International Journal of Physical Distribution & Logistics Management* [online], 19 (8), 3.

Stevenson, M. and Spring, M., 2007. Flexibility from a supply chain perspective: definition and review. *International Journal of Operations & Production Management* [online], 27 (7), 685–713.

Stewart, D. W., Shamdasani, P. N. and Rook, D. W., 2007. *Focus Groups* [online]. 2nd edition. Thousand Oaks: Sage.

Stuart, I., McCutcheon, D., Handfield, R., McLachlin, R. and Samson, D., 2002. Effective case research in operations management: a process perspective. *Journal of Operations Management* [online], 20 (5), 419–433.

SWISSMEM, 2021. *Jahreszahlen 2019* (English title: *Yearly Figures 2019)* [online]. Available from: https://panorama.swissmem.ch/de/jahreszahlen-2019.html [Accessed 30 January 2022].

Taleb, N. N., 2007. *The black swan: the impact of the highly improbable* [online]. New York: Random House.

Tang, C. S., 2006a. Perspectives in supply chain risk management. *International Journal of Production Economics* [online], 103 (2), 451–488.

Tang, C. S., 2006b. Robust strategies for mitigating supply chain disruptions. *International Journal of Logistics: Research & Applications* [online], 9 (1), 33–45.

Tate, W. L., Ellram, L. M. and Bals, L., 2019. Supply chain finance: solutions for financial sustainability, risk management and resilience in the supply chain. *In:* Tate, W. L., Bals, L. and Ellram, L. M., eds. *Supply Chain Finance: Risk Management, Resilience and Supplier Management.* London: Kogan Page, 1–12.

Thomson, R. and Arney, E., 2015. *Managing People: A Practical Guide for Front-line Managers* [online]. 4th edition. Hoboken: Taylor and Francis.

Tracey, M. and Smith-Doerflein, K. A., 2001. Supply chain management: what training professionals need to know. *Industrial and Commercial Training* [online], 33 (3), 99–104.

Trkman, P., de Oliveira, M. P. V. and McCormack, K., 2016. Value-oriented supply chain risk management: you get what you expect. *Industrial Management & Data Systems* [online], 116 (5), 1061–1083.

Tuckman, B. W. and Jensen, M. A. C., 1977. Stages of Small-Group Development Revisited. *Group & Organization Studies* [online], 2 (4), 419–427.

Tummala, R. and Schoenherr, T., 2011. Assessing and managing risks using the Supply Chain Risk Management Process (SCRMP). *Supply Chain Management: An International Journal* [online], 16 (6), 474–483.

Ulrich, P., 2011. *Corporate Governance in mittelständischen Familienunternehmen* (English title: *Corporate Governance in medium-sized family businesses)* [online]. Wiesbaden: Gabler.

van Ruler, B., 2019. Agile communication evaluation and measurement. *Journal of Communication Management* [online], 23 (3), 265–280.

van Teijlingen, E. and Pitchforth, E., 2006. Focus group research in family planning and reproductive health care. *The journal of family planning and reproductive health care /*

Faculty of Family Planning & Reproductive Health Care, Royal College of Obstetricians & Gynaecologists [online], 32 (1), 30–32.

van Teijlingen, E. and Hundley, V., 2002. The importance of pilot studies. *Nursing Standard* [online], 16 (40), 33–36.

VDMA, 2021. *Mechanical engineering—figures and charts 2021.* VDMA: Frankfurt. Available from: https://www.vdma.org/documents/34570/6128644/Maschinenbau%20in%20Zahl%20und%20Bild%202021.pdf/43a31467-dc91-1bd9-41ee-97413c4e769d [Accessed 30 January 2022]

Venohr, B., Fear, J. and Witt, A., 2015. Best of German Mittelstand—The world market leaders. *In:* Venohr, B. and Langenscheidt, F., eds. *The Best of German Mittelstand: The World Market Leaders.* [online]. Köln: Deutsche Standards Editionen, 5–22.

Venohr, B. and Lang, T., 2014. Deutsche Weltmarktführer (English title: German world market leaders). *In:* Langenscheidt, F. and Venohr, B., eds. *Lexikon der deutschen Weltmarktführer.* Offenbach: GABAL-Verlag, 22–23.

Venohr, B. and Meyer, K. E., 2007. *The German Miracle Keeps Running: How Germany's Hidden Champions Stay Ahead in the Global* [online]. Berlin: Berlin School of Economics, Institute of Management. Report Number 30.

Verbano, C. and Venturini, K., 2013. Managing Risks in SMEs: A Literature Review and Research Agenda. *Journal of Technology Management & Innovation* [online], 8, 186–197.

Villena, V. H., 2019. The Missing Link? The Strategic Role of Procurement in Building Sustainable Supply Networks. *Production and Operations Management* [online], 28 (5), 1149–1172.

Villena, V. H. and Gioia, D. A., 2018. On the riskiness of lower-tier suppliers: Managing sustainability in supply networks. *Journal of Operations Management* [online], 64, 65–87.

Vincent, L. H., Bharadwaj, S. and Challagalla, G., 2004. *Does Innovation Mediate Firm Performance?: A Meta-Analysis of Determinants and Consequences of Organizational Innovation* [online]. Atlanta, Georgia: Georgia Institute of Technology. Report Number 2004–01.

Voudouris, I., Lioukas, S., Makridakis, S. and Spanos, Y., 2000. Greek hidden champions: lessons from small, little-known firms in Greece. *European Management Journal* [online], 18 (6), 663–674.

Vousinas, G. L., 2019. Supply chain finance: definition, modern aspects and research challenges ahead. *In:* Tate, W. L., Bals, L. and Ellram, L. M., eds. *Supply Chain Finance: Risk Management, Resilience and Supplier Management.* London: Kogan Page, 63–95.

Wagner, S. M., 2010. Supplier traits for better customer firm innovation performance. *Industrial Marketing Management* [online], 39 (7), 1139–1149.

Wagner, S. M. and Bode, C., 2006. An Empirical Investigation into Supply Chain Vulnerability. *Journal of Purchasing and Supply Management* [online], 12 (6), 301–312.

Wagner, S. M. and Bode, C., 2008. An Empirical Examination of Supply Chain Performance Along Several Dimensions of Risk. *Journal of Business Logistics* [online], 29 (1), 307–325.

Walker, B., Bovet, D. and Martha, J., 2000. Unlocking the Supply Chain to Build Competitive Advantage. *The International Journal of Logistics Management* [online], 11 (2), 1–8.

Walravens, A. and Filipović, N., 2013. Three Bivalent Performance Factors of Hidden Champions: Ownership, Organizational Culture and Organizational Governance. *In:* McKiernan, P. and Purg, D., eds. *Hidden Champions in CEE and Turkey: Carving Out a Global Niche* [online]. Berlin: Springer, 37–52.

Wang, C., Fergusson, C., Perry, D. and Antony, J., 2008. A conceptual case-based model for knowledge sharing among supply chain members. *Business Process Management Journal* [online], 14 (2), 147–165.

Weiß, Y. and Kamm, S., 2020. Upskilling for "Purchasing 4.0". *In:* Schupp, F. and Wöhner, H., eds. *The Nature of Purchasing: Insights from Research and Practice* [online]. Cham: Springer, 77–88.

Werner, H., 2020. *Supply Chain Management: Grundlagen, Strategien, Instrumente und Controlling* (English title: *Supply Chain Management: Fundamentals, Strategies, Instruments and Controlling)* [online]. 7th edition. Wiesbaden: Springer Gabler.

Wieczorrek, S., Grunewald, M. and Spengler, T. S., 2019. Supplier development considering interdependencies in the inbound logistics of the automotive industry *In:* Bode, C., Bogaschewsky, R., Eßig, M., Lasch, R. and Stölzle, W., eds. *Supply Management Research* [online]. Wiesbaden: Springer, 99–116.

Wiedenmann, M. and Größler, A., 2021. Supply risk identification in manufacturing supply networks. *The International Journal of Logistics Management* [online], 32 (2), 650–672.

Wieland, A. and Wallenburg, C. M., 2013. The influence of relational competencies on supply chain resilience: a relational view. *International Journal of Physical Distribution & Logistics Management* [online], 43 (4), 300–320.

Wiengarten, F., Humphreys, P., Gimenez, C. and McIvor, R., 2016. Risk, risk management practices, and the success of supply chain integration. *International Journal of Production Economics*, 171, 361–370.

Witt, A., 2015. *Global Hidden Champions: The Internationalisation Paths, Entry Modes and Underlying Competitive Advantages of Germany's and Britain's Global 'Top Three' Niche Players*. Thesis (PhD). University of Edinburgh. Available from: https://era.ed.ac.uk/handle/1842/26039 [Accessed 30 January 2022].

Witt, A. and Carr, C., 2013. A Critical Review of Hidden Champions and Emerging Research Findings on Their International Strategies and Orientations. *In:* Cook, G. and Johns, J., eds. *The Changing Geography of International Business*. London: Palgrave Macmillan, 95–113.

Wölfer, K., 2010. Are Family-Owned Businesses Better Innovators? *In:* Gerybadze, A., Hommel, U., Reiners, H. W. and Thomaschewski, D., eds. *Innovation and International Corporate Growth* [online]. Berlin: Springer, 393–415.

Wong, C. Y. and Karia, N., 2010. Explaining the competitive advantage of logistics service providers: A resource-based view approach. *International Journal of Production Economics* [online], 128 (1), 51–67.

Xu, L. and Beamon, B. M., 2006. Supply Chain Coordination and Cooperation Mechanisms: An Attribute-Based Approach. *Journal of Supply Chain Management* [online], 42 (1), 4–12.

Yan, T. and Dooley, K. J., 2013. Communication intensity, goal congruence, and uncertainty in buyer–supplier new product development. *Journal of Operations Management* [online], 31 (7–8), 523–542.

Yin, R. K., 1984. *Case study research: design and methods*. Beverly Hills: Sage.

Yoon, B., 2013. How do hidden champions differ from normal small and medium enterprises (SMEs) in innovation activities? *Journal of Applied Sciences Research* [online], 9 (13), 6257–6263.

Zhao, L. and Huchzermeier, A., 2018. *Supply Chain Finance: Integrating Operations and Finance in Global Supply Chains* [online]. Cham: Springer.

Zhexembayeva, N., 2013. In Search of Sustainable Business in Central and Eastern Europe. *In:* McKiernan, P. and Purg, D., eds. *Hidden Champions in CEE and Turkey: Carving Out a Global Niche* [online]. Berlin: Springer, 81–93.

Ziegenbein, A., 2007. *Supply Chain Risiken: Identifikation, Bewertung und Steuerung* (English title: *Supply Chain Risks: Identification, Assessment and Management*) [online]. Zürich: vdf Hochschulverlag AG.

Zsidisin, G. A., 2003. A grounded definition of supply risk. *Journal of Purchasing and Supply Management* [online], 9 (5), 217–224.

Zsidisin, G. A., 2019. Revisiting Supply Chain Risk. *In:* Zsidisin, G. A. and Henke, M., eds. *Revisiting Supply Chain Risk* [online]. Cham: Springer, v–viii.

Zsidisin, G. A. and Henke, M., 2019. Research in Supply Chain Risk: Historical Roots and Future Perspectives. *In:* Zsidisin, G. A. and Henke, M., eds. *Revisiting Supply Chain Risk* [online]. Cham: Springer, 1–12.

Zsidisin, G. A., Petkova, B., Saunders, L. W. and Bisseling, M., 2016. Identifying and managing supply quality risk. *The International Journal of Logistics Management* [online], 27 (3), 908–930.

Zsidisin, G. A. and Ritchie, B., 2009. Supply Chain Risk Management—Developments, Issues and Challenges. *In:* Zsidisin, G. A. and Ritchie, B., eds. *Supply Chain Risk: A Handbook of Assessment, Management, and Performance* [online]. Boston: Springer, 1–12.

Zsidisin, G. A. and Smith, M. E., 2005. Managing Supply Risk with Early Supplier Involvement: A Case Study and Research Propositions. *Journal of Supply Chain Management* [online], 41 (4), 44–57.

GPSR Compliance

The European Union's (EU) General Product Safety Regulation (GPSR) is a set of rules that requires consumer products to be safe and our obligations to ensure this.

If you have any concerns about our products, you can contact us on ProductSafety@springernature.com

In case Publisher is established outside the EU, the EU authorized representative is:

Springer Nature Customer Service Center GmbH
Europaplatz 3
69115 Heidelberg, Germany

Batch number: 09538770

Printed by Printforce, the Netherlands